SECONDARY EDUCATION IN A CHANGING WORLD

Series editors: Barry M. Franklin and Gary McCulloch

Published by Palgrave Macmillan:

The Comprehensive Public High School: Historical Perspectives
By Geoffrey Sherington and Craig Campbell
(2006)

Cyril Norwood and the Ideal of Secondary Education
By Gary McCulloch
(2007)

*The Death of the Comprehensive High School?:
Historical, Contemporary, and Comparative Perspectives*
Edited by Barry M. Franklin and Gary McCulloch
(2007)

The Emergence of Holocaust Education in American Schools
By Thomas D. Fallace
(2008)

*The Standardization of American Schooling:
Linking Secondary and Higher Education, 1870–1910*
By Marc A. VanOverbeke
(2008)

*Education and Social Integration:
Comprehensive Schooling in Europe*
By Susanne Wiborg
(2009)

*Reforming New Zealand Secondary Education:
The Picot Report and the Road to Radical Reform*
By Roger Openshaw
(2009)

*Inciting Change in Secondary English Language Programs:
The Case of Cherry High School*
By Marilee Coles-Ritchie
(2009)

Curriculum, Community, and Urban School Reform
By Barry M. Franklin
(2010)

Girls' Secondary Education in the Western World:
From the 18th to the 20th Century
Edited by James C. Albisetti, Joyce Goodman, and Rebecca Rogers
(2010)

Race-Class Relations and Integration in Secondary Education:
The Case of Miller High
By Caroline Eick
(2010)

Race-Class Relations and Integration in Secondary Education

The Case of Miller High

Caroline Eick

RACE-CLASS RELATIONS AND INTEGRATION IN SECONDARY EDUCATION
Copyright © Caroline Eick, 2010.

All rights reserved.

First published in 2010 by
PALGRAVE MACMILLAN®
in the United States—a division of St. Martin's Press LLC,
175 Fifth Avenue, New York, NY 10010.

Where this book is distributed in the UK, Europe and the rest of the world, this is by Palgrave Macmillan, a division of Macmillan Publishers Limited, registered in England, company number 785998, of Houndmills, Basingstoke, Hampshire RG21 6XS.

Palgrave Macmillan is the global academic imprint of the above companies and has companies and representatives throughout the world.

Palgrave® and Macmillan® are registered trademarks in the United States, the United Kingdom, Europe and other countries.

ISBN: 978–0–230–10708–3

Library of Congress Cataloging-in-Publication Data

Eick, Caroline, 1956–
  Race-class relations and integration in secondary education : the case of Miller High / Caroline Eick.
    p. cm.—(Secondary education in a changing world series)
    ISBN 978–0–230–10708–3 (alk. paper)
   1. High school students—Maryland—Baltimore County—Social conditions—Case studies. 2. Students with social disabilities—Maryland—Baltimore County—Case studies. 3. Education, Secondary—Social aspects—Maryland—Baltimore County—Case studies. 4. School integration—Maryland—Baltimore County—History—20th century.
   5. Baltimore County (Md.)—Race relations—History—20th century.
   I. Title.

LC208.4.E35 2010
373.1809752'6—dc22   2010014348

A catalogue record of the book is available from the British Library.

Design by Newgen Imaging Systems (P) Ltd., Chennai, India.

First edition: November 2010

10 9 8 7 6 5 4 3 2 1

Printed in the United States of America.

To Emmanuel, Francesca, Joseph, and Annie

Contents

List of Tables ix

Series Editors' Foreword xi

Preface xv

Acknowledgments xix

Introduction 1

Part I The Divided Generation (1950–1969)

1 Memories of Class, Race, and Gender Divisions: Immediate Pre- and Post-Desegregation Years 17

2 Cautiously Negotiating Social Divides: A Conservative Student Body 45

Part II The Border-Crossing Generation (1970–1985)

3 Memories of Interracial Peer-Group Affiliations: Integration Years 61

4 Bridging Social Divides Through Peer-Groups: A Socially Tolerant but Politically Inactive Student Body 87

Part III The Redivided Generation (1986–2000)

5 Memories of Segregation by Class, Race, Nationality, and Religion: Destabilizing Years of Shifting Demographics 103

6	Oppositional Self-Segregation: A Student Body Sensitized to Discrimination	133
Conclusion		147

Addendum Methodology: The Transparent Historian	157
Notes	175
Bibliography	215
Index	221

Tables

A.1	Pseudonyms arranged by generations, graduation dates, and interview dates	167
A.2	Distribution by race, gender, and class (at the time of school attendance) of graduates across generations	168
A.3	Population Growth in Miller Town	174

Series Editors' Foreword

Among the educational issues affecting policy makers, public officials, and citizens in modern, democratic, and industrial societies, none has been more contentious than the role of secondary schooling. In establishing the Secondary Education in a Changing World series with Palgrave Macmillan, our intent is to provide a venue for scholars in different national settings to explore critical and controversial issues surrounding secondary education. We envision our series as a place for the airing and resolving of these controversial issues.

More than a century has elapsed since Emile Durkheim argued the importance of studying secondary education as a unity, rather than in relation to the wide range of subjects and the division of pedagogical labor of which it was composed. Only thus, he insisted, would it be possible to have the ends and aims of secondary education constantly in view. The failure to do so accounted for a great deal of difficulty with which secondary education was faced. First, it meant that secondary education was "intellectually disorientated," between "a past which is dying and a future which is still undecided," and as a result "lacks the vigor and vitality which it once possessed" (Durkheim 1938/1977, p. 8). Second, the institutions of secondary education were not understood adequately in relation to their past, which was "the soil which nourished them and gave them their present meaning, and apart from which they cannot be examined without a great deal of impoverishment and distortion" (10). And third, it was difficult for secondary school teachers, who were responsible for putting policy reforms into practice, to understand the nature of the problems and issues that prompted them.

In the early years of the twenty-first century, Durkheim's strictures still have resonance. The intellectual disorientation of secondary education is more evident than ever as it is caught up in successive waves of policy changes. The connections between the present and the past have become increasingly hard to trace and untangle. Moreover, the distance between policy makers, on the one hand, and practitioners on the other has rarely seemed as immense as it is today. The key mission of the current series of books is, in the spirit of Durkheim, to address these underlying dilemmas of secondary education and to play a part in resolving them.

In *Race-Class Relations and Integration in Secondary Education*, Caroline Eick examines student relationships in the years between 1950 and 2000 at Miller High School, a pseudonym for a comprehensive high school in Baltimore County, Maryland. At the heart of her account is the story of how three generations of Miller students interpreted their school experiences and in the process negotiated issues of class, race, gender, and nationality. The events that she describes occurred in the larger context of the struggle in the United States, the South, and Maryland to racially integrate the schools in response to the 1954 Brown v. Board of Education decision.

Hers is an oral history that focuses on the testimonies of graduates during these three generations and considers the extent to which the relationships that were built during their time in school carried over into their adult lives. She identifies these groups as the Divided Generation (1950–1969), the Border-Crossing Generation (1970–1985), and the Redivided Generation (1986–2000). Students, both high school and college, are prominently featured in much of the current scholarship on the civil rights movement and student activism. What distinguishes Eick's work is the fact that she goes further than most studies that explore the political roles of students during the struggles that occurred during these years surrounding desegregation. Her focus is on how students' situated identities, within intersections of class, race, gender, and nationality, were shaped by and in turn shaped institutional norms. It is the story of how students of different backgrounds at different times established relationships that propelled forward the alternative forces of integration and segregation.

In the book, Eick uses the outlook and skills of the oral historian to explore a place that has often been ignored in our histories of school desegregation. Her account places the meanings, beliefs, and perceptions of those participants in the foreground. It recognizes that personal memories are not simply objective descriptions of events. Rather, they are in her words, "…gendered, racialized, and class-based." They remind us that the same events may be seen differently by the various individuals and groups who participate in this or that historical event.

Eick sees her volume as offering an important contrast to another book—Gerald Grant's *The World We Created at Hamilton High* (1988). There is, she notes, an important difference between the two books. Grant's account, as she sees it, is an overly optimistic one that views the less than positive black-white student relationships that prevailed in the school that he studied at the beginning of the movement for integration as improving over time and ending in a period that Grant himself described as constituting "genuine racial equality." Her account, written almost a quarter of a century later than Grant's, tells us a less linear, progressive story. She describes demographic changes that occurred at Miller that brought into the school groups, particularly Russian Jewish immigrants, "angry black city youth," and poorer white youth, who constructed student relationships with a racist discourse. She concludes her book by considering how in different ways the struggle for integration at Miller High both advanced and retarded the ability of these three generations of high school youth to live in the globalized world that was emerging in the last half of the twentieth century and has now fully arrived in the twenty-first.

Eick's book is important in two key respects. It examines the mixed and incomplete results that secondary schools played in the struggle for school integration in the last half of the twentieth century and considers how in the midst of that struggle the American high school served to both unite and divide future generations. Second, her book personalizes the events surrounding one of the most contentious times in our recent history and brings those events into the day-to-day lives of individuals. Events that are typically described at the level of policy and programs become a compelling narrative of real people struggling around real events that affect their lives in deep and profound ways. It is a book that offers a more sober picture of the impact of efforts at desegregation and integration than other similar accounts of this period in our educational history.

Race-Class Relations and Integration in Secondary Education is the eleventh volume to be published in our series. It exemplifies well the combination of social, historical, and comparative approaches to secondary education that we have sought to emphasize throughout, and continues our focus on the interplay between the issues of race and education. As we see the trajectory of the series advancing during the next few years, our intent is to seek additional volumes that bring these issues still further to the attention of studies in secondary education.

BARRY M. FRANKLIN
GARY MCCULLOCH
SERIES CO-EDITORS

References

Durkheim, E. *The Evolution of Educational Thought: Lectures on the Formation and Development of Secondary Education in France*. London: Routledge and Kegan Paul, 1938/1977.

Grant, Gerald. *The World We Created at Hamilton High*. Cambridge: Harvard University Press, 1988.

Preface

The Supreme Court decision in *Brown v. Board of Education*, which struck down public school segregation in 1954, was, for a segment of black and white liberals, a hopeful legislation.[1] Many believed that desegregating public schools would increase economic opportunities for less-advantaged students and attune the more privileged to issues of social justice;[2] and that having practiced meeting across economic, racial, ethnic, and cultural divides during their formative years, generations of students would grow up to be adults who continued to meet across those divides; and that their acquired habits of association would translate into a citizenry who more equitably shared in the political, economic, and social life of its democracy.

However, resistance to school desegregation in the South *and* the North, by whites *and* blacks, before *and* after *Brown*, would challenge what some historians have identified as the power of an educational legislation to serve as a "catalyst for substantial changes in social relations and policies outside of school."[3] Our nation's story of school desegregation would be further complicated by the second-largest wave of immigration within a century in U.S. history.[4] By the 1990s, the meanings attached to categories of racial identity had gained considerable complexity as more diverse peoples entered a country sensitized to issues of civil rights and multiculturalism.[5] Against an imagination of greater equity and opportunity, too many immigrant children of color continued to be funneled into the lower academic tracks of our nation's public schools. Too often they were misdiagnosed as needing special education, and too often trapped in English-language learner classes, out of which they never graduated.[6] Thus at the turn of the twenty-first century, the expectations that might have been attached to *Brown v. Board* by some remain unrealized.[7]

As I write this history, 90 percent of U.S. citizens own a mere 29 percent of the wealth of this nation;[8] schools are more segregated even as the population has grown more diverse;[9] and the number of sexual, ethnic, and religious hate crimes has risen.[10] A cursory glance at our nation's economic disparities and at the racial, ethnic, and recently renewed religious tensions reveals a divided citizenry ill at ease with diversity. Still, we have primarily continued to entrust schools with the task of bringing us closer together, politically and economically. From *Brown v Board of Education* in 1954, to the Elementary and Secondary Education Act of 1965, to the Bilingual Education Act sanctioned by the Supreme Court in 1974, to the Education for All Handicapped Children Act of 1975, and to the No Child Left Behind Act of 2002, we have, over the course of the last half of the twentieth century, continually reappointed schools as our nation's equalizing forces and entrusted them with closing gaps and bridging divides.

Domestically, the story of our pluralistic democracy in the latter part of the twentieth century has been in great part the story of the relationship between our nation's schools and its citizens. Understanding ourselves as diverse peoples occupying shared social spaces at the turn of the twenty-first century is to understand the spaces that we have occupied within desegregated schools over the latter part of the twentieth century; it is to understand the generational transmission of school-acquired habits of association—in sum, to understand school over time from diverse students' points of view.

More than any U.S. institution, public schools have progressively assembled greater numbers of diverse citizens on a daily basis within confined spaces and against their preference or explicit will.[11] In particular, it is within desegregated comprehensive public high schools that the greatest numbers of students at the brink of full citizen participation have continued to meet across economic, racial, ethnic, cultural, national, and religious divides over the past half-century. Patrick Ryan suggests that "more than any other institution, the increasingly comprehensive high schools of the twentieth century redefined the social lives of American youths... [and that] our interpretation of [the comprehensive high school] should be central to our understanding of the country."[12]

Our interpretation of diverse students' relationships with one another, in an institution "as indicative of American society as any institution,"[13] should be central to our understanding of diverse peoples in the United States—particularly at a time in humanity's history when global interdependence is accelerating. How we construct relationships across differences within our own borders, how we bridge or carve divides in our own backyard, will shape how we construct our shared humanity beyond our borders. At a time when we are actively participating in the massification[14] of education across the world, while "we are neither very united nor very comfortable with our diversity,"[15] it is important to take stock of our civic relationships as they have developed since *Brown v. Board*, when we officially attempted to redefine them.

This history explores students' relationships across gender, race, ethnicity, nationality, and class divides as experienced within the architectural, academic, and extracurricular spaces of Miller High, a comprehensive high school in Baltimore County, Maryland, during a half-century marked by desegregation, civil rights movements, suburbanization and urbanization of suburbs, and the second-largest wave of immigration in the United States within a century. It is timely for two main reasons. First, and as previously mentioned, because it contributes to a reflective examination of our national capacity for associative living across census divides since desegregation, at a time when our schools are resegregating at a rapid pace. In this sense, it is a history that sheds light on how one of our important institutions has helped to both unite and divide us—in hopes that "our future may be found in the past's fugitive moments of [harmony] rather than" discord.[16]

This history is timely also because it is written when globalization is redefining the educational aims of nation-states. In addition to the dialectic tension between national democratic forces of cultural homogeneity and heterogeneity,[17] a new tension has emerged at the turn of the twenty-first century between global and local forces. Scholars have begun to suggest that within this new global/local ("glocal")[18] tension, the primary purpose of education—for both developed and developing nations—will be to help citizens adapt to multiple cultural contexts and affiliations within and beyond national borders. This history, then, by bringing into view a recent half-century of student associations in a U.S. school, during a period of great social diversification, offers possibilities for considering the capacities of the U.S. educational system to educate a citizenry at once "American" and global.

My viewpoint, in telling the history of Miller High between 1950 and 2000 from the perspectives of diverse students and as revealed through alumni oral histories, is that generational memories *simultaneously* tell of family, community, the nation, and the world—because ground-level human experiences are many daily interwoven experiences. The experience of school is also the experience of traveling to school, and of having or lacking space to do homework at home after school; of belonging to a church whose pastor is familiar with the school principal, or living in a segregated neighborhood whose elders are rarely seen on school premises; of being brought up Republican or Democrat in a conservative or liberal town; of losing sleep over terrorist threats while still having to attend school. Individual memories, nestled within generational memories and compared across generations, powerfully tell of patterns of intersections between school, social context, and citizens' lived experiences.

I share with the immortal historian Howard Zinn the conviction that "nations are not communities and have never been, [and that] the history of any country, presented as the history of a family, conceals fierce conflicts of interest."[19] Thus I chose to tell the history of one desegregated comprehensive high school from a ground-level perspective—that of students, but more important, from students' situated racial, gender, ethnic, national, and class perspectives. The world is different if one is a poor or wealthy, white or black young woman attending Miller High in the early days of desegregation—different yet from that experienced by a white or black young man, and different still across time periods as gender and race relationships change over time. I also strove not to impose preconceived affiliations between and across racial, gender, ethnic, and class differences as I collected and analyzed graduates' experiences. I strove not to essentialize *agency*, but to look for evidence of it as it came. While using categories such as "race" and "ethnicity" runs the risk of perpetuating social constructs that have no basis in science, they are the categories by which the participants in this oral historical study identified themselves and others; they are the constructs that continue to distinguish the lived experiences of U.S. citizens.

I also share with philosopher Amy Gutmann the conviction that "cultivating mutual respect entails understanding people not merely as abstractions, upon whom [we project our] own conception of what constitutes a good life, but understanding people in their own particularity, with their own lives to lead and their own conceptions of what constitutes a good life."[20] Thus in writing this history, I strove to highlight the particularities of the lives recounted by diverse alumni as they understood them. I hoped to produce a history that cultivated mutual respect even as it examined it.

This book then is a story of U.S. diversity that tells of processes of integration as experienced by three generations of citizens during their formative years at Miller High between 1950 and 2000: the Divided Generation (1950–1960), the Border-Crossing Generation (1970–1985), and the Redivided Generation (1986–2000). It explores our capacities for associative living across differences as these were nurtured or stifled within the classrooms, hallways, and sports fields of Miller High, as well as during dances, extracurricular activities, and across academic tracks. It is at once a history of youth, of diverse students, and of race, gender, and class relations in an institution that desegregated as early as 1956, and of an evolving social context marked by shifting demographics. Over this half-century, the town transformed from rural to suburban to urban-suburban; from a predominantly white middle-class town, alongside which lived a small African American community established in the nineteenth century, to a multicultural population that by the 1990s included Russian immigrants and African American youth newly arrived from city schools.

This work rests on my conviction that the evolving story of U.S. democracy, in particular as it reflects a "world lived in common with others,"[21] is in large part the evolving story of diverse youth in public schools, and that the strength and vitality of a just U.S. democracy rests in great part on how well its citizens learn to cross divides during their formative years.

Acknowledgments

I gratefully acknowledge the following historians of education who have believed in the importance of my work and who have supported my efforts in writing this book. I thank Barbara Finkelstein for generously taking the time to shepherd me through the early stages of my writing, for seeing the published work that I could not then imagine. I thank Eileen Tamura for sustaining my efforts along the way. In her presence, I found strength and resolve. I thank Kim Tolley and Nancy Beadie for articulating, with refreshing clarity, the feedback that always propelled my work forward. I thank my mentor, Philo Hutcheson, for his first-class mind and heart, and the constancy of his integrity and honesty. I thank Sevan Terzian for never hesitating to help when asked, and for providing thoughtful feedback to my early work. I would also like to thank Bob Croninger, sociologist of education and lifelong student of educational policies, who, upon reading the crudest first draft of my manuscript, saw in it a book worth teaching. I thank with all my heart my family—my husband, Joseph, sounding board and wordsmith, who also kept house and home together while I lived the writing, and my son, Emmanuel, and daughter, Francesca, who enthusiastically cheered for me. Last but not least, I thank all the women and men who freely gave their memories—the heart and matter of this book.

Introduction

This is the story told by three generations of graduates who attended Miller High, a Baltimore County comprehensive high school in Maryland, between 1950 and 2000: the Divided Generation (1950–1969), the Border-Crossing Generation (1970–1985), and the Redivided Generation (1986–2000). Each generation of graduates constructed memories of cross-group relationships quite differently, and diverse graduates within each generation told of different limitations and possibilities for negotiating class, race, gender, and nationality lines—if they were women or men, white or black, native-born or immigrant; if they had attended the upper or lower academic tracks, if they had participated in sports, and if they had been of richer or poorer economic background when attending Miller High.

Out of these graduates' recollections emerges a history of integration that extends the story of desegregation beyond struggles to secure equal representation of students across race and ethnicity. It tells of how and which desegregated school spaces translated or failed to translate over time into integrated spaces, and of the forces that sustained and sabotaged *authentic* integration—understood as the "intellectual and social engagement across racial and ethnic groups,"[1] as well as across class and gender divides. Thus this history also adds the dimension of class and gender to the story of desegregation/integration and captures a dynamic account of changing race/class/gender relationships. It is important to underscore that desegregation and integration are distinguished in this history. As suggested above, the former refers to political and legal processes to secure balanced attendance of students across race and ethnicity in schools, while the latter refers to the actual interaction between students across racial, ethnic, as well as socioeconomic groups.

A history of education of student relationships that focuses on processes of integration is timely. First, while the existence of the public desegregated school is presently threatened by the accelerated resegregation and privatization of schools, it continues to hold our best hope for uniting us across differences. By the close of the last century, the notion that authentic U.S. democracy rests on school desegregation had become an idea that more citizens defended than did not. Scholars and the U.S. public in general continued to agree more than they disagreed that "school integration is one of the most urgent educational issues today."[2] However, it remains unclear whether over time desegregated schools have indeed helped to foster authentic integration and if they have helped lessen prejudice. This history is a contribution toward our understanding of the desegregated school as it shapes student relationships, and of the conditions that might foster its capacities as an instrument of integration rather than resegregation.[3]

This history also broadens discussions on desegregation within educational research at large. As one social scientist noted, "[W]hile student achievement data are critical, schools do more than teach academic subject matter: They have a profound potential for shaping individuals and their social networks."[4] In turn, social networks play an equal if not more important role in post-school life opportunities than might achievement. Jencks found that "academic achievement does not have the kind of overwhelming impact on later occupational success that might justify making it the exclusive focus of most research on the outcome of desegregated schooling."[5] Thus a history of education that emerges from the oral historical testimonies of graduates further allows for connections to be made between school relationships of the past and post-school social networks of the present, as graduates reveal whether the interracial and cross-class relationships sealed in school translated into broadened social networks in adult life.

This history is exploratory. It is a beginning. It raises more questions than it answers, and it is certainly not an exhaustive story of all possible configurations of student relationships as they unfolded at Miller High over the latter part of the twentieth century. Nor could such a complete story be told given the many competing historical interpretations of school life and peer relationships. But in important ways, beyond its general contribution to scholarship on student lives—an underdeveloped facet of the history of education—it contributes to scholarship on processes of integration in four important ways. First, it focuses on a suburban high school situated in the orbit of a border city in the border state of Maryland, a barely explored area. Second, it extends and complicates the important story told by Gerald Grant in *The World We Created at Hamilton High*.[6] Third, it offers a framework for historical investigations of processes of integration. Fourth, it foregrounds historical protagonists' meanings, beliefs, and perceptions as historical facts rather than anecdotal evidence to corroborate or challenge established historical facts. As the works of oral historians have suggested, memories are gendered, racialized, and class based. They bring to our attention the importance of examining why different individuals and groups experience the same historical events in very different ways.[7]

The Desegregated Suburban School in Baltimore County

While scholarship has examined civil rights student activism in institutions of higher education, and more recently in high schools, these works do not examine students' unfolding relationships across social categories *within* desegregated schools over time. They tend to focus on students' participations, as black *or* white students, in the political struggles for desegregation and protests for fair treatment.[8] This work examines student relationships within a de facto desegregated school to examine processes of integration over time—a high school situated in Baltimore County, Maryland, in the environs of a border city within a border state.

Baltimore County schools desegregated as early as 1956, for reasons unique to Baltimore and Maryland as a border city and border state, respectively. Maryland's border-state status created a particular context for civil rights movements that contrasted with what Palumbos described as the "triumphalist narrative of struggle, suffering and success," which characterized the Deep South. Maryland activists "began building African American–led institutions to fight racism before a coherent national movement was born...African Americans welcomed white support, and whites generally followed black leadership."[9] Thus Baltimore city schools desegregated immediately in 1954. However, while the city school board adopted a free-choice desegregation policy given the more supportive relationships between African Americans and whites, by the 1960s white flight, later accelerated by ensuing riots following the assassination of Martin Luther King, Jr., rapidly resegregated city schools.[10]

For some third-tier Baltimore County suburbs, those farthest from the city and within rural areas, the story would be different. In the orbit of the city of Baltimore, Baltimore County also desegregated comparatively early, in 1956—but many of its de facto desegregated schools would not resegregate. In 1967, Baltimore County created the "Urban-Rural Demarcation Line," which "preserved exurban areas as farmland in the northern section of Baltimore County."[11] These third-tier suburbs also included the county's long-established black families, whose children, beginning in 1956, would attend desegregated schools. Thus situated in the northwestern corner of Baltimore County, Miller Town is a third-tier suburb that maintained its "rural" status well through the 1970s and early 1980s. At the edge of suburbia and at the edge of white flight, Miller High would remain de facto desegregated—predicated at first on a relatively small number of African American youth, sons and daughters of long-established county families whose roots extended back into the nineteenth century and who were descendents of freed slaves.[12]

By the late 1980s and throughout the 1990s, however, Miller Town and its high school would be drastically transformed from predominantly rural-suburban to urban-suburban, from primarily white and middle class to multiracial and economically diverse. Thus Miller High represents the multiracial suburban high school that has undergone the kinds of demographic changes that many parts of the United States underwent over the latter part of the twentieth century, are presently experiencing and will continue to experience in the foreseeable future. Immigration patterns, particularly since the end of the twentieth century, continue to change the face of public schools beyond urban settings, penetrating suburban schools, which for a while seemed to many middle-class whites as well as middle-class blacks, as we shall see, "safe" from inner-city strife.[13]

Building on Gerald Grant's History

To my knowledge, Grant's work is the only other U.S. educational history that examines student relations within a comprehensive high school in the second half of the twentieth century, and only across race divides. If in Grant's story "things got very bad before they got better, but they did get better," and the school went from "white power to black power to genuine racial equality,"[14] it is in great part, I argue, because Grant's story ends in the early 1980s when the country still rode the wave of civil rights movements, and before the second-largest wave of immigration within a century had flooded U.S. schools. In light of the findings in the case of Miller High, Grant's affirmation of "genuine racial equality" holds only under certain demographic conditions.

Demographics played a crucial role in shaping Miller High students' relationships with school authorities and their interactions with peers across diverse backgrounds. The comfortable interracial and even cross-class relationships (given the middle-class continuum) that students finally achieved during the 1970s were destabilized by an influx of newcomers that brought a new type of white and black person into the community. The new type of white person included the overtly racist and "anti-American" Russian Jewish immigrant youth, and the new type of black person included the overtly angry black city youth whom African American "old timers" accused of corrupting county black youth and addicting them to drugs. The result, during the 1990s, was that a segment of white native youth whom some stereotyped as rednecks, and who might have kept their racist sentiments silent during the 1970s and into the early 1980s, when they were kept in check by integrationist white peers, became visible—as overtly anti-Russian, anti-Jew, and anti-black. By the end of the century, students sported race, ethnic, and national identities in opposition to one another.

But Grant's story also suggests that while black students' political engagement and uprising distanced teachers at first from authentic engagement with students across race, it also eventually forced a school and community reckoning with the race issue. Thus Grant's story suggests that student activism plays an important role in shaping the direction of school relationships. In the case of Miller High, a school situated in a conservative town known to vote overwhelmingly Republican—a town where, as one African American alumna suggested, blacks were "familiar with getting along with white people," and where black and white Miller Town "old timers," as we shall see, preferred the town's status quo relationships for economic as well as social reasons—students remained, across the fifty-year period, politically inactive. During the most daring crossings of race/gender/class divides by students of Miller High, in the 1970s and early 1980s, students never protested against the institutional and institutionalized injustices that they recalled as graduates in their oral testimonies.

Thus informed by Grant's history, I dare to suggest that the lack of Miller High students' political engagement during the 1970s and early 1980s when they were freest to demand changes, given the backdrop of the national civil rights mood and the relatively lax school policies, diminished possibilities for continued integration into the end of the twentieth century. The influx of Russian immigrant youth, many of whom were "anti-American" and racist, and of black youth newly migrated from the city and unaccustomed to desegregated spaces and mistrustful of the white establishment, was greeted by discriminatory institutional norms that had become more rigid by the early 1990s and that exacerbated divisions of race, class, nationality, and even religion. Even as many students in the 1970s and early 1980s had created interracial and cross-class peer groups, as reported by those graduates interviewed, their lack of political engagement against the very sorting mechanisms that they remembered as unjust, coupled with discriminatory institutional practices, allowed for a contingency of black students and poorer students to remain segregated, which would be easily fueled and exacerbated by the addition of racist youth and youth who had not developed habits of association within desegregated spaces.

Thus seismic demographic shifts that brought racist and mistrustful immigrated white and migrated black youth into school spaces, and a lost opportunity for political action by the most democratic generation of Miller High, moved the school not from "white power to black power to genuine racial equality," as was the case of Hamilton High, but from "white power" to "genuine racial equality" to resegregation and aggressive balkanization.

Situated Perspectives and Cross-Group Relationships

Histories that trace processes of integration hold the promise of capturing constructions and reconstructions of categories of social identity as these are shaped within schools, and by extension, to identify more clearly the evolving relative roles played by students, teachers, administrators, and broader social forces in defining the perimeters of relationships across markers of difference. Although using categories such as "race" and "ethnicity" runs the risk of perpetuating social constructs that have no basis in science, they are the constructs that continue to distinguish the lived experiences of U.S. citizens.

The works of educational ethnographers who have been engaged in *multivocal* and *polyphonic* narratives have helped pave the way to explore student encounters across diverse backgrounds.[15] However, lacking the historical dimension that establishes analytic linkages across time periods, ethnographies fail to capture transformations and continuities in students' experiences. Thus this analysis, while not determined by it, is informed by black feminist theorist Patricia Collins' intersectionality theory.[16] The intersectional analytic approach examines the ways in which social markers of difference (race, gender, ethnicity, sexuality, generation, class, religion, and nationality) intersect to shape situated experiences. This approach at once questions the homogenized renditions of group experiences along racial, gender, and other socially constructed categories of identity, and sensitizes the researcher to hierarchies of domination within and across intersecting markers of difference.[17] It further brings to light the shifting nature of group identities. The theory grew out of black feminists' articulation of the limitations of "gender as a single analytical category,"[18] and by extension, the limitations of any single social category to capture the complexity of cross-group relations.

Informed by the theory of intersectionality and committed to an oral historical perspective that foregrounds the subjective experience of history—not to rewrite facts, but to gain a deeper understanding of how and why the same facts are reconstructed differently across historical protagonists—I first identify students' situated experiences within each generation before I analyze more deeply their cross-group relationships. Thus I bring to light the ways that, across time periods, students reinterpret the social categories of race, class, ethnicity, and nationality; the ways they cross and reconstruct those boundaries; and the ways in which they use established categories for new identity purposes.

Historical Protagonists' Beliefs Understood as Historical Facts

In this work, I foreground oral histories as primary sources that tell about meaning, belief, and perception, and not as reflections that corroborate, disprove, or expand on historical events. Alessandro Portelli elegantly captures the possibilities of oral historical inquiry when he writes: "[T]he first thing that makes oral history different is that it tells us less about *events* than about their *meaning*." Portelli boldly asserts that "what informants believe *is* indeed a historical *fact* (that is, the fact that they believe it), as much as what really happened."[19] The intent in this historical analysis was primarily to locate *different perspectives* across intersecting social categories of identities and to capture *meanings* ascribed to relationships in school as historical events in their own right, rather than investigate a single past occurrence. Thus "difference," not "volume," mattered. The goal was not to produce a "collective" memory of Miller High students. Undeniably, such an endeavor would have required "volume," but more important, would have proven futile given the many situated perspectives that compete for the historical meanings of an institution whose population is not homogeneous.[20] While archival research preceded, accompanied, and followed oral historical interviews, and descriptive statistics were derived,[21] the time periods emerged first from narrators' voices; archival research confirmed or corroborated patterns in oral historical testimonies, not the other way around.

Because the focus was on perceptions and meanings ascribed to school relationships by alumni of different backgrounds, the "truth status of individual stories became less important than the value they [were] trying to support in the telling."[22] This work then is also an exploration of the possibilities of historical inquiry to trace evolving perceptions of social groups from multiple perspectives, research having shown that perceptions rather than facts often shape cross-group relationships.[23]

* * *

What This History Is Not

It is important to underscore that this history is not a history of youth subcultures. When youth subcultures come into view, it is because they reflect graduates' experiences of student relationships across categories of race, class, gender, nationality, and/or religion. For example, through a few of the testimonies of the graduates of the 1990s, the punks and goths very briefly come into view. The relevance of punks and goths for this history lies in their being white suburban youth as opposed to an interracial youth subculture. Another example is the prep identity, which is significant for this history because it is the one student identity consistently associated across time periods with upper economic status and superior consumer power. It is a category of identity that by the end of the century included wealthy youth across races and ethnicities. Finally, the jocks, musicians, and other peer groups of the 1970s are significant for this history in that they are the interracial, cross-class, and cross-gender groups that reorganize student relationships toward greater integration and lessened segregation.

The dual focus of this history is to capture students' relationships with peers and teachers from their situated perspectives and to explore the levels of integration across race, class, and other census categories, across time—the comprehensive high school having been assigned the task to bridge the divides of class at its inception, and to bridge the divides of race with *Brown v. Board*, as I explored in some depth in the preface.

Time Period and the Story

The second half of the twentieth century is a particularly fecund epoch within which to investigate youth relationships in high schools because it begins with the greatest expansion of lawfully sanctioned student rights (*Brown v. Board of Education* and *Tinker v. Des Moines*),[24] only to end with almost unprecedented restrictions on students' actions and civil rights (zero-tolerance policies).[25] It is a period that captures a momentous and frenetic cycle in the history of the high school and the lives of its students. High school attendance became a universal expectation, effectively bounding the experience of adolescence with the experience of school, even as the high school itself increasingly struggled for legitimacy.[26] This period has been the era of youth and of the high school, the institution through which all U.S. citizens, since the 1950s, have been expected to live their adolescence.[27]

Also within this arc of time, the real and symbolic power of *Brown v. Board of Education* would be diluted. By the turn of this century, scholarly works began noting a nationwide trend toward segregation and inequality,[28] with some historians tracing it back to the 1970s.[29] Other works began revealing the effect on suburban schools of what has come to be known as the "urbanization of suburbs," whereby more affluent city dwellers move to the suburbs seeking the good life and escaping from poor city schools and crime. The result, as some researchers have pointed out, is that city escapees do not find opportunities for a better future for their progeny, but instead an extension of urban life into the suburb. Then, too, while the Supreme Court decision in *Brown v. Board of Education* originally spoke directly to the black and white divide in the United States, by the end of the twentieth and beginning of the twenty-first century, the landscape of diversity had expanded to include not only people of races beyond black and white, but also people who shared one race but came from very different cultural backgrounds and spoke languages unintelligible to one another. In some schools, more than a hundred different languages could be heard spoken between parents and children. By the end of the century, the meanings attached to racial categories of identity had gained considerable complexity.

Thus not only did I want to trace students' evolving relationships within the high school over almost a half century under *Brown v. Board of Education*, as graduates would reveal these from their situated positions relative to each other and school authorities, but I also wanted to know if subsequent generations of students either transformed or reproduced social prejudice,, and if certain conditions or time periods were more propitious for one or the other. I wondered whether alumni recollections would confirm or disprove what Carnoy and Levin have pointed out, namely that "schools continue to provide Americans with a social experience that is markedly more egalitarian and more open to free choice and possibilities of self-realization than anything that is available to them in the realm of work."[30]

What follows is an account of the evolving nature of students' relationships with each other over the past five decades of the twentieth century as they unfolded for three different groups of Miller High graduates, which I call generations: the Divided Generation (1956–1969), the Border-Crossing Generation (1970–1985); and the Redivided Generation (1985–2000). As mentioned earlier, a complete story of all possible configurations of student relationships as they unfolded at Miller High over the latter part of the twentieth century cannot be told given the heterogeneous student population and the many intersecting situated positions. But the echoed memories of student relationships as reported by the interviewed graduates across race, class, gender, and other divides, as well as memories collected through informal interviews over a two-year period (2003–2005), suggest that enough Miller High graduates perceived cross-group relationships similarly within certain time periods, as well as differently across time periods, to constitute generations. Moreover, groups congealed that articulated in similar language those similarities out of which emerged time periods, which were further corroborated by broader demographic changes in Miller Town. Bound by shared memories of institutional practices, shared demographic configurations, and shared articulations (in use of vocabulary and cultural references) of cross-group relationships, the oral historical testimonies sketched the perimeters of generations. Within each generation, however, nuanced stories emerged from situated perspectives of, for instance, young black middle-class women when compared to those of white middle-class women, working-class men, and so on.

Recollections revealed that authentic integration at Miller High had been the experience of too few students across generations. Interracial friendships that lasted beyond school years were mostly forged during the 1970s and early 1980s, an anomaly in this fifty-year history. A propitious convergence of social forces sustained the Border-Crossing Generation: namely, a broadly shared middle-class status across racial divides; a population of Miller High black students who were mostly children of long-established families in the region and a minority; a student body well accustomed to desegregated institutional settings; and a national consciousness of civil rights that stirred the imagination of an otherwise conservative student population in a conservative town.

Recollections further revealed that cross-class friendships within and across racial divides were rare. For the Divided Generation of the 1950s and 1960s, and for the Redivided Generation of the late 1980s and 1990s, they occurred mostly in upper academic tracks and a few extracurricular activities, most notably in sports. These spaces were the least populated and most insular—spaces for the chosen few. Poorer students' possibilities for participation in upper academic tracks or involvement in after-school extracurricular activities, which proved central to developing relationships with peers across diverse backgrounds, were restricted. Those students of lesser means who could join the upper academic tracks or play in organized sports increased their chances of forging friendships across class and race divides. Thus within elite spaces, encounters across differences were more likely to be friendly. Within the more populous general and vocational tracks, on the other hand, they were more likely to be unfriendly.

Indeed, the general and vocational tracks remained overall divisive spaces, incubators of racial violence. More important, animosities were both reproduced and intensified within these tracks. In the confined spaces of Miller High, students could not avoid each other by dispersing into alternate realities as they might have in the larger community. The cafeteria also remained a place where students continued to segregate as they sought comfort in the familiar and where school authorities consistently left them to fend for themselves. During the 1990s, when demographic shifts brought into school perimeters Russian immigrant and black city youth newly arrived from segregated schools, youth who were mistrustful of the school system and of each other, and who sought refuge in familiar ethnic, racial, religious, and economic relationships,[31] spaces such as the cafeteria only exacerbated students' segregating tendencies.[32]

However, within this overarching account that tells of greater segregation than integration, many nuanced stories complicate and deepen the story. For example, young black and white women in the lower academic tracks in the early days of desegregation of the late 1950s and 1960s experienced more conciliatory racial encounters when compared to those of their male counterparts, and further when compared to those of young black and white women in the lower academic tracks in the 1990s, when they engaged in racial fights. Other stories revealed how an African American youth of means in the vocational track in the early 1960s would be considered "poor" by white peers of lesser means who racialized his economic status. The many nuanced stories strewn across this history testify to the slippery nature of an endeavor such as this one that seeks to build claims on shifting perspectives.

How Miller High students of diverse backgrounds negotiated, organized, and carved relational spaces for themselves, with each other, and with school authorities, and how they integrated or segregated is the story of this history. It is a history that identifies the dynamic interplay between demographics, institutionalized tracking,[33] disciplinary policies, and the power of social class and gender, across time, in loosening or reinforcing students' self-segregating tendencies across racial lines. Many more voices might have recreated a new set of credible experiences that might have expanded or challenged the stories recovered here. But for the graduates interviewed, these were lived experiences, and as such, historic. Then, too, and as suggested earlier, the many nuanced stories of this history resist an authoritative statement about how things were.

Still, I offer claims at the levels of structure and agency. From a structural perspective, I argue that Miller High, born of the progressive era and the social efficiency movement, and which remained effectively unchanged in its architecture and sorting organization since the turn of the twentieth century, could not, by virtue of its classist tracking, create integrated spaces for the majority of its students. Following *Brown*, Miller High black youth entered a high school world of institutionalized classism. Paradoxically, the problem was not so much that Miller High continued to be socially reproductive—since it was also the space within which diverse students, who might never have crossed paths in the broader community, continued to meet—but that it actually exacerbated social tensions for those of lesser means across race, gender, and national identities within its lower tracks, and particularly so under the stresses of major demographic shifts that substantially diversified the student population. Youth of lesser means and/or with lesser prospects of continuing their education into college segregated the most in the lower tracks. Moreover, the greater integration across race and ethnicity within upper tracks was, until the end of the century, predicated on majority white youth representation, suggesting that "integration" within upper-academic tracks was a de facto assimilation of all nonwhite youth and white youth of lesser means into a white middle- to upper-middle-class culture. Further research should examine when integration becomes assimilation rather than the "intellectual and social engagement" of youth across racial, ethnic, class, and gender divides.

Then, too, there is the case of sports. The institutionalized competitiveness of high school sports at Miller High restricted the participation of many students since only those who could compete to win participated. Thus, although sports fields would remain integrated across race and ethnicity, they would also remain elite spaces. More research is needed to examine why so many black Miller High youth were consistently accepted into competitive sports, but so many more talented black youth were never recruited, as this history suggests. Did white coaches apply laxer participation criteria for white youth and more demanding criteria for black youth? Were they striving to keep racial representation tilted toward majority white representation?

From the student agency point of view, I argue that the self-segregating tendencies of Miller High students, when confronted with new social configurations, as was the case for the Divided Generation in the early years of desegregation and for the Redivided Generation of the 1990s, were further exacerbated by distant teachers and punitive disciplinary measures. During the early years of desegregation, many black and poor white youth were met with indifference or outright rejection. During the 1990s, youth fought back and retreated deeper into their racial, ethnic, and national identities under the pressures of zero-tolerance policies.

Broadly and very generally, this analysis suggests that Miller High students of the second half of the twentieth century best bridged social divides in spaces that held high status in the school and that were least populated. It strongly suggests what seems obvious: when young people were respected, dignified, and allowed to explore associations within smaller, more intimate school spaces, they tended to unite rather than divide and to establish friendships across divides that continued into adulthood. Conversely, when young people were considered lowest on the totem pole, ignored, herded into overcrowded spaces, and made to be school visitors through work-release programs, they tended to retreat into groups of familiar backgrounds, to stereotype and fight with those who were different, and to continue to segregate into adulthood.

Parts and Chapters

The analysis that follows proceeds in three parts. Part I, the Divided Generation, is developed in two chapters. In chapter one, "Memories of Class, Gender, and Race Divides," I examine the situated perspectives of Miller High students as young men or women, black or white, richer or poorer—perspectives that proved significant to graduates who attended between 1950 and 1969. In chapter two, I examine the meanings attributed by graduates to the social categories by which they identified their place and that of others at Miller High, and I analyze the spaces, real and imagined, within which Miller High students of the Divided Generation attempted to cross boundaries of race, class, and gender, and within which they cautiously explored alternate identities with peers and teachers.

Part II, the Border-Crossing Generation, is also developed in two chapters. In chapter three, "Memories of Interracial Peer-Group Affiliations," I examine how students reorganized their identities around shared affinities that trumped categories of race, class, and gender, as revealed by graduates who attended between 1970 and 1985. In chapter four, as in chapter two, I examine the meanings attributed by graduates of this generation to the social categories by which they identified their place and that of others at Miller High, and I analyze how they navigated between the various interracial, cross-class, and cross-gender peer groups.

Finally, part III, the Redivided Generation, is also developed following the organization of the previous two parts. In chapter five, "Memories of Segregation by Class, Race, Nationality, and Religion," I examine yet another reorganization of student identities according to dimensions suggested by the chapter's title and as revealed by recollections of those who attended between 1986 and 2000. In chapter six, again as in chapters two and four, I examine the meanings attributed by graduates of this generation to the social categories by which they identified their place and that of others, and I analyze the cultural complexity as well as the rigidity that student segregation acquired by the end of the twentieth century compared with that experienced by the Divided Generation of the 1950s and 1960s.

I follow the body of analysis in the Conclusion, in which I review the social and institutional forces that sabotaged and sustained authentic integration for Miller High students over the second half of the twentieth century, further discuss the claims forwarded earlier, and consider whether indeed Miller High, either within or across periods, offered experiences "markedly more egalitarian and more open to free choice and possibilities of self-realization than anything that [was] available to them in the realm of work."

Finally, in the Addendum, "Methodology: The Transparent Historian," I discuss the voices that could not be recorded and the theoretical and practical challenges presented by *the doing* of oral history.

Part I

The Divided Generation (1950–1969)

Chapter 1

Memories of Class, Race, and Gender Divisions: Immediate Pre- and Post-Desegregation Years (1950–1969)

In the 1950s and 1960s, Miller High[1] students attended a school that had been twice renamed and relocated, and rebuilt several times.[2] Its history extended into the early decades of the twentieth century and was punctuated by landmark investments, one of which in the 1930s transformed Miller Town's older 1914 high school building into the "most modern and best equipped educational plant,"[3] to be modernized yet again in 1960 as the population grew. Throughout its transmutations, it remained a "Main Street" high school, an integral part, both structurally and culturally, of Miller Town, where many of its white middle-class teachers and administrators lived and students could easily cross paths with school authorities on streets, in grocery stores, and in churches. One of the oldest high schools in Baltimore County, Miller High of the 1950s and 1960s was also the only one within a twenty-mile radius, a lone educational establishment in the middle of farm country. One alumnus recalled: "There was a farm behind the high school, and every now and then the cows would get out in the fields...you had to watch out where you were running."[4]

Miller High students in the 1950s inherited an institution with more than one hundred years of history deeply rooted in the white community of Miller Town.[5] This was a high school with long-established rituals. Students both then and today looked forward to "the first day of school...[when] the teacher who had the most tenure rang the bell."[6]

The equally deep roots of Miller Town's African American community, which extend back to the nineteenth century, ran parallel to those of their white neighbors. Generations of black citizens of Miller Town owned small businesses, barbershops, and shoe-repair stores; trained racehorses for wealthy landowners in the area; drove trucks; and provided domestic work for affluent white households.[7] While they lived mostly clustered along Brand Avenue, a road within easy walking distance of the town's high school building, Miller Town's African Americans sent their teenagers on hour-and-a-half-long bus trips to and from the all-black high school in the nearby city of Towson until 1956, when Baltimore County desegregated its public schools.

In the 1950s and 1960s, when "suburbs replaced cities as the fastest-growing residential sector,"[8] Miller Town was still a predominantly rural town and would continue to be so through the 1970s. As one alumnus stated: "We were very middle-class white rural...Oh yeah, we used to get called farmers."[9] Immediately following desegregation, and into the late 1970s, only a handful of African American students would attend Miller High, all of them children of long-established black families of Miller Town whose livelihoods were deeply interconnected with those of their white neighbors and vice versa.

Thus, throughout the 1950s and 1960s, Miller High served an overwhelmingly white population within a predominantly rural setting. Accordingly, until the late 1960s it offered courses in horticulture and animal husbandry, as well as the Future Farmers of America club. It also offered extracurricular activities that reflected the era's broader national politics of concern for international peace—the foreign exchange program and the United Nations Youth Organization—and others such as Students for the American Way, Future Teachers of America, Future Business Leaders of America, and Future Nurses of America, which reflected local demographics, economics, and conservative politics. Miller High provided Miller Town's labor force of farmers, clerks, teachers, factory workers, and businesspeople;[10] educated the college-bound elite; and served as a center for performances, celebrations, fairs, and expositions.[11]

Early Years of Desegregation

When black students began attending Miller High in 1956, they entered a world that had been comfortably inhabited by a majority of white middle-class students and their white middle-class elderly teachers, many who had careers that extended back to the early decades of the century. White middle-class girls in particular, whom teachers and administrators favored, suffused Miller High with their sensibilities and personalities throughout the 1950s and 1960s. To understand Miller High students' differently situated experiences during this early desegregation period, one must first understand the high school world that white middle-class girls participated in creating. It is by contrast to the white female middle-class normative experiences that white male students', black students', and poor students' experiences come into view most dramatically.

White Middle-Class Young Women

Young white middle-class women[12] were at home and at ease at Miller High, an institution that for them held porous boundaries and that they easily entered and exited. These girls moved in a high school world of fluid roles and organizational structures that sustained easy border crossings between school, community, and family life. Student, teacher, and staff could swiftly change to errand girl, grandmotherly figure, or parental authority. As alumna Dorothy Kaufman remembered, permission slips were readily available:

> Mrs. Smith would invariably say, "You want to get out of your study?" Sure. So she'd write a note, and we'd run up to the principal's office and she'd send us to the drugstore to get whatever she needed. I walked right by my father's grocery store and I waived and he waived back...We'd go buy for her and she'd give us money to have a Coke while we were there.[13]

They attended an institution drenched with ancestral memories, where generations of older siblings, mothers and fathers, and even grandparents, lived their high school lives before them. They frequented a high school with a principal, Mr. Lancaster, a "very popular man, both with kids and school,"[14] whom they loved and respected. They lived in a world where their elderly teachers might send their favorite girls on errands to the drugstore in the middle of school days for headache pills.

White middle-class adolescent girls at Miller High in the 1950s and 1960s set the pace and flavor of the school academically, socially, and representationally. Academically, in that they worked harder than the boys, filled the ranks of national honors societies,[15] and never contested assignments, thus raising the bar for achievement. Moreover, they often took care of their male counterparts by offering to do their homework, a habit that would persist into the early 1980s.[16] As one alumnus remembered:

> I had a couple of girls in the classroom that would either take notes for me or sometimes they would give me what they had written as their assignment. And I would copy it or I would change it.[17]

The girls set the tone socially in that they were May queens, homecoming queens, and Miss Senior High; took over the planning of dances; filled the majority of student council seats; and could "almost get away with murder."[18] Alumna Alice Web recalled how easily she duped her teacher by simply being of "respectable background":

> I was looked upon as a good kid, and so I could get away with murder...people whose parents were professionals, or who seemed more respectable [narrator gestures quotation marks as she speaks the word "respectable"]...I would miss the bus on purpose to walk to school 'cause then I'd get there at about ten o'clock and I'd just say, "Oh, Mrs. Reece, I missed the bus again." "Alright," [she'd say] and she'd write me an excuse, you know. And she never said, "You sure miss the bus a lot." You know, I was very rude because nobody stopped me.[19]

Girls also set the tone representationally in that they overwhelmingly determined the content and layout of the yearbook (*The Key*) and forged the opinions of the *Miller Chronicle*.[20] Thus they extolled pride in their school; presented authority figures with reverence; and portrayed students fast at work in classrooms, posing for sports or club group pictures facing the camera straight on and in orderly rows. Linda Moss, a 1969 graduate, captured these young women's involvement:

> We were the doers. We were the people that were on the teams, we were the people that were putting out the yearbook, were the people doing that, you know, class officers...Maybe we were snobs, or elitist, or something like that, but I don't remember feeling that way.[21]

These young women shared an easy transition into a high school where they continued longtime friendships established since elementary school and where authority figures were familiar extensions of parental care.[22] They were also the most likely to receive awards on behalf of the school, such as the then-prestigious Freedoms Foundation Award, "organized in 1949 to further the ideals of American democracy...to create a love of freedom...and to build an understanding of the Constitution and The Bill of Rights."[23]

While in most every way, as alumni reported, white middle-class females made Miller High their home; they were considered smarter than their male counterparts, privileged in the responsibilities awarded them, highly visible in almost every social aspect of the school life, on good terms with teachers, and even allowed to "get away with murder." They differed significantly in what they could do with their high school education and in what a high school diploma meant to them, given the track they attended.

For college-bound middle-class women throughout this period, Miller High was a stepping-stone. Alice Web, graduate of the class of 1954, remembered: "I never questioned that I would go to college...No question I would go. Anybody who had the money could go."[24] Graduates in the 1950s Dorothy Kaufman and Alice Web, and 1960s graduates Sherry Parson, Betty Land, and Linda Moss, alumnae whose fathers could afford college tuition, took it for granted that they would attend institutions of higher education. While Dorothy applied herself and sought good grades, Alice, bored and restless, did minimal work and misbehaved, never doubting that college awaited her.

> If I was interested in the class, then I would pay attention and behave. If there was nothing going on that interested me, I was very apt to get into trouble...There was no pressure to get good grades...never worried about it [about getting into college].[25]

After graduation however, alumnae went on to become teachers. The historically documented trend of women enrolling in teacher programs since the nineteenth century continued throughout the 1950s and 1960s in Miller Town.

For lower middle-class adolescent girls with no allotted budget for college, the academic track was not an option. While they may have excelled academically and widely participated in extracurricular activities, not having money for college meant choosing the next best thing: the commercial track. By default they prepared to enter the business world as secretaries or took art classes.[26]

Whether they attended the academic track to eventually become teachers or the commercial track to be hired as secretaries, looking the part of the middle to upper-middle class was of essence. It was because Judy Law was able to look the part and hide her lower-class status that she could be at home at Miller High. She explained how lucky she had been:

> I was very lucky in one respect, because I was considered to be a cute girl...I was Miss Senior High...My family didn't have any money and they didn't have any status. I was lucky in the respect that I had a grandmother that sewed for women in the Valley. I had gorgeous clothes...My family was very poor and they cared about the image they put out in the community. They wanted me to be clean and well dressed.[27]

Although she knew from the start that college was out of reach, Judy, safely camouflaged in her middle-class look, went about being smart, working hard, and getting involved in a myriad of school activities. She elaborated:

> I was an honor roll student...[in the business track]. My best girlfriend and I were selected [by the principal] to accept an award for our school.

> It was called the Freedom's Foundation Award. And we went to Valley Forge with our principal. I have a news clipping of that. I was involved in student council, I was involved in sports, and I was involved in the Future Business Leaders of America. As I got older I was president of that. Because of my background, coming from a family that didn't have money, I was not put on that track [referring to academic track]. I always knew that my parents didn't have the money to send me to college.[28]

In general, high school life for middle-class and middle-class-looking young women was an empowering experience, whether academically or socially. It provided the social and virtual spaces to practice life roles, some with direct correspondence to adult roles that awaited them, and others with no theaters available for them to play out in the immediate and broader cultures of their times. Thus they practiced creating dances to socialize with their male counterparts, actively participated with them in developing "going steady" dating patterns,[29] and immersed themselves in organizing fund-raisers and a multitude of social events. Such social skills, practiced throughout their high school years, could then easily be put to use by these young women as future middle-class housewives. Presiding over the Future Business Leaders of America club or being editor of the *Miller Chronicle,* however, held no equivalent positions in the business communities of the time for these young women. Soon after graduation, some of them felt cheated by such disparities in high school-to-world correspondence, as if they had been lied to.[30] Miller High was the molding ground for a way of life for these young women. They were supported by a network of older middle-class women teachers, who, at the end of their careers, looked the other way when their charges transgressed, often elicited grand-daughterly behavior from them, and spoiled them with privileges and extra money for treats. A network of social players kept things in place for them. As Dorothy explained: "You had all your role models; you had parents, and you had your teachers. If you went to church, you had those people...it sort of gobbled you up, you know, took you in."[31] Judy shared how she "had a lot of encouragement, especially from the older women teachers,"[32] and all alumnae echoed Nora's testimony when she said: "Our teachers took interest in us."[33]

White middle-class and middle-class-looking adolescent girls were at home at Miller High, and made it their home. They saturated its clubs, extracurricular activities, and classrooms with their well-groomed appearances, accompanied by semi-docile, semi-mischievous behavior. They knew the rules well, and knew their teachers' tolerances well, as one knows one's parents' limitations and weaknesses. Their favored position at Miller

High was further corroborated by males' points of view. Alumni Nat Right and Bud Land recalled:

> The girls were treated much better than boys. Girls seemed to get the good side of everything. You had an argument, the boy was always wrong. I don't care what happened. I don't think I'm being prejudiced. That's what they showed me. The girl's always right no matter what.[34]
>
> From a male standpoint, you know, they [faculty and administration] were always easier on the girls.[35]

White middle-class girls navigated the emotional life of school with ease, fluently read the hidden cultural messages of allowable transgressions, and dexterously manipulated the boundaries of acceptable behavior across authority figures and school contexts. Dorothy captured the level of comfort that these young women enjoyed in the presence of authority figures when she fondly recalled when she and her future husband skipped school to see a basketball game, only to meet with Mr. Lancaster, who had stepped out of his principal role to be the ice cream vendor:

> [The] championship basketball game came up between Miller, and I think it was in Kenwood... So, my husband and I were dating at that point and we decided we were going to go to the game... So we took off and there's Mr. Lancaster [the principal] standing there, and he says, "Now, Jim and Dorothy, enjoy the game. I'll see you tomorrow in detention. By the way, I'm selling ice cream." [Doris laughs heartily.][36]

White middle-class girls judged teachers' professional skills apart from their personalities and confidently assessed their abilities regardless of the teachers' dispositions.[37] Whether they judged their teachers as "good" or "bad," these young women could count on them to pay attention to and help them. However, while they shared the world of appearance, comfortable relationships with authority figures, and possibilities for self-expression and leadership roles within Miller High, they did not necessarily share the meanings of their attendance.

For 1950s graduate Judy Law, whose parents were poor, the daily escape from isolation and boredom of country life, opportunity to practice being someone else, and welcoming atmosphere created by teachers and administrators made going to school an experience of deliverance,[38] and the high school diploma itself was the social stamp of approval of her indisputable rights to autonomy.[39] For non-college-bound women like Judy, the high school diploma not only attested to one's official crossing into adult status, but also brought honor to the young adult's family, elevating her in the eyes of the community. Parents whose child graduated from high school were vindicated for past failings and simultaneously acknowledged for the shared success in their daughter's achievements.[40]

For Dorothy and Alice, who were college-bound, high school was a place to have fun on the way to college. While Dorothy worked hard for "two to three hours in the evening, and homework on the weekend,"[41] and assiduously practiced for music recitals and theater plays, she enjoyed every moment of it. This was when she met her future husband, with whom she participated in theater productions. Alice, on the other hand, took advantage of her advantages and got into mischief. She confided: "I'd say come on, we can do it...so we'd sneak off and go swimming, or sneak off and go to the drugstore and get a soda."[42] For 1960s college-bound graduate Sherry, high school was as a "rite of passage that you have to go through," one that she thoroughly enjoyed and during which she "gained good friendships...learned [her] academics [and] to get along with others, leadership skills, those kind of things."[43] For Betty, it was her group of friends that made high school a good experience, and for Linda, it was the support of the faculty and learning that convinced her she could "succeed at stuff."[44]

Whether or not they were serious students, they shared fun times and were at ease with themselves and their environment. These were confident young women for whom "it was a very nice status quo community," a "really a great place."[45] They lived as elite students in the rarified air of the top of the hierarchical totem pole. Dorothy explained:

> In those days...if you didn't go academic, it almost infringed upon your intelligence...People who go to college are it...Of course that's not true, but then.[46]

For these girls, school spirit soared at Miller High,[47] and graduation was "pomp and circumstance," an acknowledgement of their indisputable appurtenance to Miller High.[48]

Poor and Working-Class Young White Women and Men

By contrast, some white girls did not feel "tied to" Miller High. Girls who got pregnant, girls with learning disabilities, and poor girls often dropped out. Indirect descriptions of their lives through others' remembered accounts sketched vague impressions of timid young girls who were lost, alone, and with no place to fit in.[49] Pregnant girls were gossiped about, and whether they disappeared or graduated, they lived a peripheral life in the memories of solidly integrated white girls. Usually more than one girl got pregnant by senior year. The boys who impregnated the girls remained anonymous in narrators' remembered accounts, accounts that well capture the already historically and sociologically explored onus on women of that era for getting pregnant.[50]

If one couldn't dress the part and appear middle-class trim and proper, one was not included. Teachers' support and encouragement, generously lavished on girls who played their parts, were withheld from those who "might not dress well...[and] smelled bad."[51] Familiarity, intimacy, and trust, easily extended to the "right-looking" girls, were denied to the visibly poor. Through Alice Web's middle-class perspective, the very poor came to view in this way:

> The poor kids...who might not...dress well and so on...smelled bad...there was a kind of snobbery about the poor kids. They [the poor] were people who lived in shacks. The teachers didn't try to bring them out...people would talk about poor kids having bugs and that kind of stuff...when they'd do the hair to check the lice, it was always the poor kids who would have the lice, and they would smell funny.[52]

Among those shunned by students and teachers were the poorer white farm boys. Within the predominantly rural community of Miller Town, farm boys represented a significant portion of the student population. However, while Miller High offered agricultural classes and the Future Farmers of America Club, raised its own pigs, and sponsored the yearly fall fair where livestock could be displayed, it was predominantly young men whose fathers were landowners who took advantage of these activities. Boys who would not inherit big farms and for whom training to show prized cattle or choosing the best fertilizers for crops was not useful visited school between jobs. While middle-class boys whose fathers could afford college wore "khakis and...drove their daddy's cars,"[53] James Dean look-alikes like alumnus Robert Heart wore jeans and white undershirts with rolled-up sleeves in which they tucked in their cigarettes, and they drove their bikes to school until they mustered enough money to rebuild old beat-up cars. High school was not a priority for Robert Heart and his friends. They could get through it by just showing up. Robert explained: "I could care less about school...I said, I'm not going to college. So I don't care what I do. I just need to get out, okay?...Just show up."[54]

Robert's recollections of his high school experiences and of those of his friends reveal that the lives of poor farm boys revolved around jobs that made ends meet.[55] There was barely any time to catch up on sleep, let alone do any schoolwork.[56] For these young men, school was a system to endure, retaliate against, reject, feel embarrassed about, or humiliated by. At their best, they suffered teachers; at their worst, they fought them. As Robert put it: "I had a teacher that kicked me out of his class three times... And he had a Volkswagen... And we just picked the Volkswagen [up] and put it on the front steps of the school."[57] Pride gave some a Hollywood-like bravado as they punched out the villain teachers and sometimes walked off into the sunset, vindicated, to become successful entrepreneurs, never to return to the old stomping grounds. Remembering one of his more aggressive friends, Robert said:

> The phys. ed. teacher... kept picking on us... and he screams at us, and you know, take three more. And I said, 'I can't do it. I'm going to work'... So we just walked off the field. So he came over and he got in one of the guy's face, and the guy smacked him and broke his nose, walked right out of the school, and kept on going. He never came back... he is a very successful businessman right now.[58]

These young men studied in the general track and took shop courses. It was in shop class that Robert and many of his friends found a school authority who spoke their language, a male authority figure who simultaneously appealed to their rough edge and nurtured their talents.[59] Many learned early on to fight and fend for themselves. Robert Heart, whose father's death forced his mother to be on welfare for seven years while she cleaned churches, and who remembered coming home from school "and having to take the shoes off and go barefooted because you only had one pair of shoes,"[60] recalled:

> I used to show up at the bus stop, and there was all these seniors... I was always the guy being picked on... they would throw my books over the fence and make me climb over the fence and get them [just before the bus would come]... I was the littlest guy... My mother said to me, "... you have to learn to fight for yourself."[61]

Poor farm boys learned to retaliate when they were picked on—and they were picked on a lot. When peers called them "rednecks and hillbilly,"[62] they "would end up knocking the living hell out of them."[63] Major fights usually happened "on the front lawn of the school."[64] When reflected through the perspectives of those who called some poor farm boys "hillbillies," these were bigoted youth who hated blacks. Sherry Parson remembered:

> Hoo. The fights that went on. There was a bar named Franky's and it was kind of unsavory... and it usually was between a black person and a hillbilly—that's what we called them back then.[65]

African American alumnus Nat Right identified the white student who had knocked his teeth out, as we shall see later in greater detail, as a "hillbilly":

> But honest, when I became an adult, I probably would have hurt him. I thank God our paths did not cross...I forgave him...Not being accepted by the whites or blacks, see, that's where they're coming from. We [the blacks] had a hard time being accepted by the whites. They [the hillbillies] had a double stance. See what I'm saying? The hillbillies were supposed to be the lower-class whites. I can't think of any other way to say it. That was the way it was.[66]

Alumni across time periods alluded to the racism of some of these white boys. They were boys who grew up tough and became sensitized to class differences.[67] Still, although some dropped out—feeling more at ease in the workplace, where they spent most of their time and reaped needed monetary rewards, incentives that found no equivalent match in school—many did graduate. For those who stayed, attendance and diplomas were about beating the system and proving one's endurance. Robert said: "I was determined to be there and to finish the system, even if I had to beat the system somehow."[68] Paradoxically, Miller High was also a place that anchored Robert's life, no matter how unappealing, not unlike a military service.

> I mean, what would I be, who would I be if I didn't go to some kind of school that gave you some regimentation. You know, something that you *had* [narrator's vocal emphasis] to do, something that forced you to do. Someplace that took you to an end.[69]

School was a daily battle not only with some peers, but also with teachers, "teachers who wouldn't bend for [a student]."[70] Unlike the correct-looking white adolescent girls who often wrapped teachers around their fingers, white farm boys felt "picked on" by them.

> They would call on you four times in the class, knowing that you didn't know any of the first questions. And he would continue calling on you for the rest of the day. Trying to make a fool out of you...they were picking on us [farm boys].[71]

For them, teachers' personalities and attitudes toward students were inextricably linked to their efficacy as teachers. While middle-class white girls could have a personal rapport with teachers and still consider their teaching ineffective, working-class farm boys expected a teacher to be at once personable, compassionate, and understanding, as well as be able to deliver differentiated instruction. These young men did not separate a teacher's personal and professional personas.[72] They attended Miller High not as recipients of instruction, but as critical consumers of its deliverance. For them teachers were not familiar people one was to please or receive praise from, but paid agents of a system that owed them.[73]

Other white males of Miller High who were not headed for college, who were not part of the academic track, and who were not necessarily farm boys lived their high school career on the edge of their classroom seats, ready to spring out of them when the local fire department alarm went off. Not one teacher stopped these young men from exiting, sometimes through school windows, when the alarm in town rang. There was a tacit understanding that they would be indispensable to the town in the near future as volunteer firefighters, and therefore should be excused without question from school in the immediate present. Still other white males missed school during hunting season, a yearly ritual inherited from earlier times, and so thoroughly integrated into the cultural expectations of the community that, again, no penalties were exacted for missing school on those days. These were often boys who became the local firemen, mechanics, and construction workers.

Miller High School was not a place that working-class white adolescent men made their own; rather it was a place they visited, a place where middle-class and middle-class-looking white girls ruled, and where in caretaking fashion the girls often did homework for their male counterparts. Robert summed up the white boy–girl divide when he recalled how teachers "expected less of the boys... [and] they weren't going to get anything out of [the boys]"[74]

In general, poorer white male students lived on the edge of Miller High School boundaries. They often missed classes to work, to serve as firefighters in training, or to go hunting with their fathers. While some had it harder than others, they were all expected to find work in the community after graduation. Overwhelmingly, too, they filled the general track; as one alumnus recalled: "[I]t was mostly the guys that were in the general with shop."[75] In the 1950s and 1960s, it was the track considered by students who did not attend it to be a dumping ground for the unable and unwilling student. Bud Land, who had been college bound, described that the general track was meant "for those who were just going to go out and work as laborers. They were the slower kids and the ones that didn't really care about what they were doing."[76] Bud's testimony suggests that middle-class students at Miller High during that period had integrated the notion that those attending the general track were naturally inferior. Ironically, Bud's own memories revealed a young man disinterested in the school curriculum, one who could not wait to get out of Miller High.[77]

White Middle-Class Young Men

Bud Land and his friends were part of a group of white male students who lived through the academic track with ease and nonchalance. Although college bound, they felt no pressure to prove themselves. Bud shared how he "didn't always do what [he] was supposed to."[78] But he knew how far he could deviate from the "conformist" behavior he identified in his white female counterparts, among them his future wife and her friends,[79] before he turned into a seeming "radical." By the very end of the 1960s, a group of white male students, as Bud remembered, were "really out there... they were the ones that were experimenting with some of the drugs... [had] long hair... really long hair... the radical group."[80] These radicals were part of the general track. Just as other alumni recalled, Bud also remembered that tracking determined one's associations "all the time. That was one thing about school. You just moved through with the same group."[81]

Young White Men across Socioeconomic Backgrounds

Bud, a white male college-bound graduate of the late 1960s who put little effort into his studies, shared with Robert Heart, a poor white male graduate of the late 1950s who barely attended school, a perception of teachers as representatives of a system that ultimately did not side with students. Bud recalled: "It was always us versus them—teachers and administrators. I don't remember the teacher getting in trouble for anything. The teacher was the administration, and then there were the students."[82] White males of the 1950s and 1960s, whether rich enough to go to college or not, in the academic track or the general track, as recollections suggested, identified themselves as students *against* teachers. White middle-class girls identified themselves as students *with* teachers. In either case, these young people spoke of their place in school in reference to teachers, and as such, reflected on their place in high school as *students*. Testimonies suggested that white males, such as Robert and Bud, who were poor students regardless of their socioeconomic backgrounds, and who perceived a strong bias on behalf of school authorities in favor of girls, tended to develop more or less antagonistic relationships with teachers. For them, high school was irrelevant to life, and its teachers were out of touch with students.

African American Middle-Class Young Women

For Miller Town black students entering Miller High in 1956, however, school was anything but irrelevant. The first generation of black students who attended Miller High in the first five years or so immediately following Baltimore County's 1956 implementation of desegregation reported absorbing a swell of suspicion, and in some cases, outright animosity. The first years of desegregation at Miller High were particularly difficult for black female students.

In 1956, African American alumnae Annie Cole and Doris Wright entered an institution forced by law to teach them. Unlike their white female counterparts, who experienced an easy transition into high school, Annie and Doris recalled living through a traumatic border crossing, the pains of which scarred their memories. Abruptly uprooted from the all-black school they had been attending, they entered an institution with a foreign tradition and alien history. It was "like being in another world."[83] Annie vividly recalled the first day at Miller High:

> So this particular day...the principal at Washington High said, "Everyone that rides the Miller Town bus from grades seven to eleven, I want you to go to the auditorium," and he said we were to get on the bus and don't ask questions...we did come back to Miller Town and the bus turned into Miller High...so we all went to the auditorium and he [Mr. Lancaster] said, "This is the school you will be attending." My heart went, "Oh, God." I was terrified.[84]

This fearful experience sealed in them longings for the comforts of *their* traditions and a sense of home as they knew it. Annie and Doris remembered missing their old school.

> I missed Washington so much. Oh, boy. The programs we had, and the different things we did, and sometimes, we even could walk and get a hamburger, and do different things. But it wasn't like that here [at Miller High]...It was more stiff [at Miller High].[85]
>
> I think I would have rather graduated from Washington because I had been there from seventh grade.[86]

Unlike the fluid ways in which white adolescent girls traveled Miller High's social and academic boundaries, African American adolescent girls, as Annie reported, were restricted, their behaviors tightly bound to performance in a no-nonsense, businesslike fashion: "learn and behave."[87] To them, Miller High felt strict and harsh, devoid of warmth, emotionally disconnected from the familiar. Its boundaries were rigid and hard to penetrate—so hard that, as Annie's recollection suggested, black moms did not visit:

> My mom used to bring lunches and hot dogs...and she would make cake and bring it to us [at the old school]...it wasn't that kind of atmosphere at Miller High. It was strictly a learning thing. You just had to learn and behave.[88]

Annie walked the hallways fearfully, assailed with spit and unsavory handling by white male students. She spent her energies protecting her body and her feelings, all the while having "to learn and behave." Going to an all-white school took every ounce of effort, took a lot, and finally took too much. Annie dropped out after about a year and a half. She remembered the assaults:

> There was a whole lot of name-calling and spitting on you...They had stairways...you'd go all the way down and you could feel the spit drop on your head and people [boys] would feel your behind, and you know, my mom used to starch and iron my blouse...and it would be ink all over...I couldn't stand it...You know, I left. I graduated from home.[89]

Now and then, however, as Doris recalled, a black girl might be tentatively greeted by a white girl timidly reaching out across the racial divide, more in symbolic gesture than actual friendly contact:

> She [white adolescent female student] used to wave at me [Doris makes the gesture of hand waving timidly with elbow glued to the side of the body], because she was afraid to really talk to me because some of the [white] kids didn't want her to talk to us.[90]

It was within classrooms and during physical education that black teenage girls, as alumni recalled, found safer spaces, moments of intimacies with some teachers, and moments of equanimity with other teachers whose standards for and behaviors toward students remained equal, regardless of color.[91] Subject instruction and teachers' personalities were two separate experiences for Annie and Doris, just as they were for white middle-class or middle-class-looking girls, as discussed earlier. Whether or not a teacher was liked did not determine these students' assessments of the teacher's instructional abilities. However, unlike the experiences of white-middle-class girls for whom neither help nor instruction were ever denied, regardless of the quality of instruction or personalities involved, access to teachers was often denied to black middle-class young women, just as it was to the visibly poor white women. For Annie, relationships with teachers were uneven. Within the engineered racial encounter[92] of the time, and within an all-white school with a long history of being "nicely status quo,"[93] an African American adolescent girl might be denied instruction altogether, not just given poor instruction, because she was black:

> I remember my home economics teacher. She wouldn't help me with my work and I just felt out of place...But would say, "I'm busy now. You'll have to wait until I get to you." But that day never came. If you knew an answer to a question...and if you wanted to know something from the teacher, and you raised your hand to participate in the class, you never got called on. Never.[94]

Doris and Annie moved through an unpredictable, unstable, contradictory, and somewhat surrealistic world inside the walls of Miller High, where peers might assault them on their way to classes or reach out through secret, timid hand signals. Where teachers might support and receive their confidences or withhold instructional assistance.

They were not at home at Miller High, a shifting *Alice in Wonderland* sort of place. Doris and Annie described:

> Going to an all-white school is like being in another world...No, it was just like you were in another world. [Doris shakes her head in disbelief and wonderment].[95]
>
> Some of the teachers understood, and some of them didn't, and some of them just didn't give a damn...just a very uncomfortable situation. It was like everybody had to do this thing because the law said so...And that's the way it was. It wasn't because we're happy. No...it was very hard.[96]

When teachers did nothing to correct and punish white students' verbal or other assaults on them, Doris actively practiced blocking these students' behaviors from her awareness, while Annie sought the help of the principal.

> I don't remember the cafeteria. Some of the things you just blocked out. I don't remember the cafeteria at all...I can do it, I can block out everything around me, and I think I must have been doing it at an early age. Hate to tell you, they'd sit there, talk to me, and I didn't hear. I can block things.[97]
>
> I used to go to the principal. And he used to say to me, "You're doing very well, you're very intelligent." But that wasn't what I was there for, you know..."What you need to do is just try to overcome all of this...It's only a handful of you all, and it's two hundred and some white kids," he said. "It's just going to be hard for me to control this. So what you have to do is just try to overlook it all." So I did for a long while. I couldn't, I just couldn't. I left.[98]

As Annie recalled, the principal, handcuffed by "some of the families that had been in the community helping the school"[99] for years, didn't do much publicly to help change school-wide racist attitudes. Individually, however, he praised black students' work and made public their achievements through honor rolls.[100]

Middle-class African American young women, in the early days of desegregation, unlike their white middle-class female peers who outshined their male counterparts academically and socially, lived in the shadow of their male counterparts, who protected them, chided them, and attracted greater attention. Annie remembered going to her track-star brother for help regarding a bully white boy:

> Sam DiPaglia. I remember him. He used to bother me all the time. He used to punch me in my back, you know...I went and told my brother. My brother said, "Stop whining, don't whine, just ignore him." But he kept on doing it. So I said, "You've got to help me." So he [her brother] went and told him and he [Sam DiPaglia] finally stopped...all I had to do was tell [my brother] who was bothering me and he would go to that person and [that person] would stop...My brother, he got along fine with everybody. They loved him because he was the track star.[101]

For Annie and Doris, Miller High was an imposed exile from the comfortable and familiar, a place where one daily braved humiliations and faced potential bully assaults to the body. They wearily navigated the emotional life of school, spending most of their affective energies in self-protection. Moments spent with a friendly principal, teacher, or peer were far fewer when compared to white middle-class and middle-class-looking girls' easy interactions. None of the black young women were enrolled in the academic track during the early years of desegregation, although they would begin to do so by the mid-1960s.[102] Moreover, none of them, as was the case for many of the poorer white students, participated in after-school activities.[103] Some worked after school, and others were required to head home directly as a measure of protection.[104] While they shared the anxieties wrought by tenuous relationships with authority figures and peers, and together walked a foreign school world that they did not visit after school hours, they differed in how much of the Miller school life they wanted and in their willingness to forge ahead. For Annie, the emotional price exacted for attending Miller High, regardless of her academic achievements, was too high, and the diploma was best acquired in the safety of her home. Thus "dropping out," for Annie, was not a reflection on her poor academic standing, but rather a strategy for protecting her emotional integrity.[105] The energies were spread thin of the black female honor roll student who had "to learn and behave" while her dignity was daily assailed in school, and who was also the babysitter at home and an after-school laborer actively participating in providing financially for her family. Something had to give.[106] For Doris, who did not have to work but who still could not attend after-school activities, there were aspects of Miller High worth exploring should one have had the time. Albeit an alien and at times dangerous world, it seemed to offer possibilities of making her dreams come true and opportunities of which to take advantage:

> Because I went only to eleventh and twelfth grade...As far as I was concerned, we didn't really have enough time at Miller...A lot of the things that were taught there [at Miller High], we didn't have at Washington in Towson [narrator's previous all-black high school]...we were learning something different...I was interested in going into the corporate world...I joined FBLA [Future Business Leaders of America].[107]

For those black young women who graduated throughout the latter years of the 1950s—and nearly all graduated[108]—the graduation itself was as filled with conflicting emotions as had been their attendance at Miller High. A young black girl could be proud of her accomplishments and bring pride to her family yet simultaneously feel disconnected from the school she was graduating from and eager to get out of it. Doris remembered:

> Yes, I was [proud of the achievement]. [But] I just wanted to get out of school. I really wasn't that crazy about school...it was very important, because my parents stressed the fact that you will graduate, you will go to college. But I didn't go to college.[109]

While middle-class African American girls in the early years of desegregation shared feelings of alienation and were denied access to academic tracks like their white female counterparts of lesser means, regardless of their level of participation in the life of Miller High, the experiences of athletic young black males who proved themselves academically were quite different.

Middle-Class African American Young Men

Several African American young men attending Miller High immediately following desegregation were highly esteemed by teachers and peers alike. They were stars on the playing fields in sports competitions and honor roll students in upper academic tracks. Norman Good, a graduate of the class of 1959, was such a young man. For him, Miller High was a place where teachers helped students succeed and peers cheered for one another:

> In many instances I did better in mathematics and English than some of the white students. And they accepted me as one of them when they saw that I could compete.[110]

However, proving oneself at first took some finessing. Norman described how he gingerly offered answers to questions for which he often was the only one to know the answers:

> When there would be questions with regards to grammar, or whatever, and I would know the answer, I would, you know, look around, you understand, being in a situation where I was the only black, I would look around to see if any of my classmates would have the answer, which they didn't. I would shyly raise my hand, and the teacher, my English teacher...a rather elderly lady at that time, she was soon to retire...she would, wondering if I knew the answer or not, she would say, "Norman?" and I would answer it. And this would happen a number of times...then my classmates [would think], "This guy, he knows all the answers."[111]

Norman and young black men like him were the first to set *firsts* for black adolescent students at Miller High and in the county. Norman was the first black student of Miller High to be honored with the Millership Award, and the first black student to win first place in the Baltimore County science fair. Norman and Annie Cole's brother were among the first black students to excel in sports competitions representing Miller High.[112]

Unlike African American adolescent girls who suffered at the hands of bullies and felt "out of place," black adolescent males who competed well academically and in sports felt included by their white peers and never experienced any direct violence.

> In terms of picking on other students, you know, there wasn't. And in terms of violence, there wasn't any violence per se...I can't say any that I personally witnessed. I would hear about it. But I think that the school officials, not the least of which the students, the students of good will, they handled it, handled it well.[113]

Norman's recollections of distant rumors of fights were echoed by many graduates in upper academic tracks within and across time periods, further underscoring the relentlessly insular effect of the upper academic track.

Norman Good, Willie James, and Annie Cole's brother Jimmy Cole[114] were students with reputations as sports stars and scholars. As mentioned earlier, they often played the roles of mediators for younger siblings or other black students whose integration into the social fabric of the school was not so smooth. These young black men participated in after-school activities such as track, high-jumping, basketball, and soccer.[115] Some, like Norman, attributed their scholarly successes to the good training they had received at the all-black high school,[116] and they were out to prove to anyone who might have doubted that they were as good as their white counterparts in the classroom as well as on the field. These were highly motivated students: "I would continue to be the best possible student that I could...I competed in a good-spirited way for grades."[117]

Thus Norman understood his place at Miller High as that of a student there to do the work of a student. Within the rarified air of the upper academic track populated overwhelmingly by white students with means, and on the sports fields after school hours, middle-class black young men forged long-lasting bonds with their white male peers, bonds that extended beyond graduation.[118] Experiencing Miller High from the vantage point of its upper academic track gave black adolescent men opportunities to mingle with whites whose fathers owned stores and whose families held high profiles within the community. This, in turn, augmented their status in the white community.

> I was in a class with an individual whose family owned one of two drugstores. They were prominently known...There was another individual whose dad owned...a jewelry store...and an individual whom my mother worked for doing day work. He made a comment to his mother about how well I dressed and how good a student I was.[119]

For Norman, getting into the upper academic track, however, required much more than good grades on transcripts. Above all, it required firm parental insistence. Immediately following desegregation, black students' transcripts were transferred from the all-black school in Towson, where these students had acquired the confidence of scholarly success built on track records of accomplishments. Norman spoke of his eagerness to prove himself at Miller High and the insistence it took on his parents' part to have him attend the upper academic track:

> I looked forward to the experience [of attending a predominantly white school]. I never doubted my ability...When I arrived at Miller, they had a three-track program...and although they had my transcript of the grades that I had at Washington High, the counselor, I recall, said to my parents, "Well, we think Norman should be able to do very well in the general program." My parents said, "No, we want him in the academic program."...Counselor had some concern about whether I would be able to compete. "Well, okay, fine, what about the commercial track?" [the counselor asked]. My parents said, "No, the academic program." I did very well...I graduated in the top ten of my graduating class.[120]

While counselors were apt to acquiesce to black parents' requests to place their successful students into academic tracks, immediately following desegregation they were neither recommending these students for white higher education institutions nor informing them of options beyond black colleges.[121] Thus graduation held mixed feelings for the college-bound African American young men of Miller High. They had set precedents, brought pride to their families, established collegial relationships with their college-bound white counterparts, but their choices for higher education were restricted, limiting in turn the continuation of their freshly established relationships with "successful" whites. Norman Good's description captures the complexity of sentiments regarding and meanings attributed to graduation and the high school diploma, as he longed to continue his education at a white university:

> Yes. A day of excitement, you know. It was a happy day and in some respects a sad day, because after having been with, at least the ones in the academic program, being with them for three years and knowing that they were going off to universities in different parts of the country, and not knowing whether you would see them...you know, that was the sad part...My classmates, many had gotten scholarships to the University of Delaware, to Towson University, one of them I think the University of Pennsylvania...I would have loved to have gone to one of those schools. Based on my academic record, I would have been able to do so...But anyway, the counselor had just mentioned to me about Morgan State University. I had gotten a scholarship to Morgan, an academic scholarship...I would really have liked to go to University of Maryland.[122]

For other African American male students, attending even historically black colleges was not an option. Middle-class African American young men who did not make it into the academic track moved in more treacherous circles. They were more likely to be attacked and sucked into fights. For them, Miller High was a "black and white" thing.[123] By law, they were desegregated, but in their experience, they were "still segregated."[124] Nat Right, a 1963 graduate, recalled that integration was "something we *had* to do."[125] His recollections sketched one painfully vivid scene immediately following desegregation when Nat was assaulted by a white peer, identified, as mentioned earlier, as a "hillbilly."

> Next thing I know, he hit me in the mouth. I lost six teeth and six had to be pulled. Over the years I have gotten over it. I'll leave it at that. He is dead and gone and my life is changed. I'm a Christian now. He can be forgiven...I was out of school myself for about four months. I had to have gum surgery...That is why I failed that year. Because of the fact I wasn't in school. Because of that.[126]

While college-bound black young men fought for their integration through academic competitiveness, non-college-bound black young men, along with non-college-bound black young women of this early desegregation period, fought for their *bodies* to be allowed and respected within the same high school space that white bodies occupied. For them, fighting was not a rumor but a visceral encounter. Annie's recollection of an infamous Halloween party underscored how black bodies were denied access:

> There was a whole lot of fighting, and I remember this particular time when they had the first Halloween party, they had a dance. And I said, we can go to that now...We almost had a riot...And I didn't understand what was going on, but my brothers did. They said, no, [the white boys] they're trying to start trouble...so it finally escalated. Somebody hit somebody. Somebody called somebody that bad name... And then, that was it...I remember my brother hit somebody in the head with a stick. It just wasn't a good thing.[127]

While black male students enrolled in the upper academic track only heard about skirmishes and thought that school officials handled incidents adequately, other black male students actually took and delivered the blows and found no justice from school authorities, only implorations to try "not to take everything to heart...to ignore most of it."[128] Similarly, poorer white male adolescents, who did not participate in sports and were not enrolled in the upper academic track, were more likely to find themselves embroiled in fights, but they also directed their aggressions toward male teachers.

Thus the lower general track, attended mostly by males, primarily poorer white males (many of whom, as earlier suggested, were racist farm boys) and black middle- and working-class males, became the site of racial strife. All the while, Mr. Lancaster, judged the fairest of principals by whites and blacks alike and beloved by his community, felt powerless to provide any redress for the racist violence against his black students whom privately he praised. However, by the mid-1960s, life at Miller High became more bearable for young black women and men, as graduates' recollections suggested.

Settling into Desegregation

For Dotty Moris, the first day of high school was a far cry from the experiences of Doris Right and Annie Cole in the early days of desegregation. Dotty remembered her first day in 1963 as uneventful: "But as I can remember, it was just a normal day. We walked. So it was quite a few of us that walked together... and then we met with others, the same kids that went to Miller Elementary."[129] Dotty began her high school career alongside neighborhood friends who had graduated with her from primary school. Following an easy transition into a new level of schooling, she soon expanded her circle of friends across racial boundaries:

> And then I met other people, new friends... we just became best friends. So you had an opportunity to meet new people that you just didn't even know that were in your area... I met a lot of white friends.[130]

Dotty had "a good relationship"[131] with her teachers and got involved in many school activities early on, pursing interests that she had begun exploring in middle school.[132] In Dotty's story, things were no longer just "black and white" as they had been for Annie and Doris. Dotty, who was not spending her energies protecting herself from physical and emotional assaults and who had already practiced being a student in a desegregated elementary school, remembered relationships with peers less in terms of skin pigmentation and more in terms of friendly or unfriendly behaviors across races.

As Dotty stated, while there were white students who still gave black girls the cold shoulder, there were outright bully black students one strove to avoid.[133] *Bad* girls and boys, according to Dotty, were homebred, and student behavior was a direct reflection of familial upbringing, whether white or black. The black and white polarization of the early desegregation period was further challenged in the mid to late 1960s with the beginnings of interracial couplings at Miller High.[134] Less than a decade after the county had implemented desegregation, and within fewer than two generations of graduates since the Halloween dance incident, a lone interracial couple in a still very rural Miller Town dated and married. Their relationship forged within the world of high school.

For Dotty, Miller High in the 1960s was a great school that upheld high standards of achievement and excelled in county sports competitions.[135] In her experience, the teachers who "were older... they were patient but yet firm in their teaching," and equal with all:

> I don't think there was a difference because I was black and you were white. That attention wasn't given [by teachers] or that it was more demanding here and not there. I didn't see that.[136]

According to Dotty, if one didn't succeed in a subject matter, whether one was white or black, it wasn't because of the teacher, but because one hadn't put in the effort required of a student:

> If you didn't get it, it was because you didn't want to learn or you didn't want to take the time. So in that respect, I can say that it [academic standard] was demanding.[137]

In Dotty's memory, teachers were good disciplinarians who "would come out and really talk to the kids and break up fights"; they were hands-on authority figures who took care of their charges. Mr. Lancaster, who retired in 1964 after presiding over Miller High for fifteen years, was remembered fondly by Dotty, as he had been by all alumni, white and black.

While Dotty's memories tell of her feeling more a part of her high school than Doris or Annie ever did or could, and of her forging long-lasting friendships with black and white peers and participating in extracurricular activities, she still only spent two years at Miller High before leaving at the end of her tenth grade:

> My girlfriend got pregnant, I got pregnant, and we just ended up finishing, I finished my tenth grade. I ended up going to finish in the city and so did she; we went together and we finished our last year and graduated...It was more for embarrassment. If the times were like today, and people accepted more, I mean, it's nothing now. Then, you were stereotyped. 'Oh, she's having a baby, she's not married,' you know what I mean...Going back to school with friends that you were close to, you know, you don't want to do that.[138]

The tender subject of an adolescent's pregnancy and Dotty's genuine vulnerability and courage in sharing her story settled my dilemma of further probing into the nature of her embarrassment vis-à-vis friends. Dotty's rewarding beginning at Miller High leaves one imagining her high school career as a successful one. By the late 1960s, as yearbook data and informal interviews suggested, and as the career of Annie Cole's sister attests to, young black women attending Miller High would continue to succeed academically. Annie's sister, who graduated from Miller High in the early 1970s, went on to become a nurse.[139] By the end of the 1960s, life at Miller High for middle-class black young women had become more comfortable.

The lives of young middle-class black men who were stars on the playing fields in sports competitions continued relatively unchanged throughout the 1960s. They continued to hold a privileged status among peers as athletes, and at the very end of the 1960s, when football was introduced, as jocks, and they were most likely by virtue of this status to develop friendships with their white male peers.[140] Yearbooks reveal very few black adolescent males enrolled in the academic track throughout the 1960s. Those who were, however, were also jocks. Referred to by narrators who graduated in the 1950s as "athletes," jocks would continue to hold, throughout the second half of the twentieth century, a privileged position in the recollections and imaginations of Miller High alumni. Jock life was a place where black and white men collaborated.

For those middle-class young black men who neither attended the academic track nor were part of the jocks' crowd, life got a little easier. Nat Right, a 1963 graduate, remembered a Miller High where "after the first couple of years [immediately following desegregation], things started to mellow out, [and students] got down to being students."[141] Nat's traumatic experience when a white boy hit him in the mouth and he missed a full year of school while undergoing gum surgery did not stop him from pursuing his high school education. Still, while Nat thought eventually things calmed down between whites and blacks (especially following the Halloween dance incident), and while he remembered Principal Lancaster as "probably one of the fairest ones there,"[142] his overall experience continued to be one of discrimination.

> There was some discrimination...[teachers] ignored you if you raised your hand...we were not supposed to have knowledge. It was perceived back then that blacks were inferior...There was always somebody in your class who couldn't stand you because of your color...You had them in every class.

As Nat recalled, Miller High was still a fairly segregated place:

> Blacks tended to hang out with blacks, and whites tended to hang out with whites, and girls tended to hang out together, and boys tended to hang out together.[143]

By 1967, when African American alumnus Burt Sadden graduated, the transition into high school for young black men was much easier. Burt recalled that "it was pretty easy"[144] going from middle school, where he had already forged many friendships, to high school, where he continued those friendships. By the time Burt attended Miller High, and as Nat's testimonies suggested, things had changed, and blacks and whites had gotten used to working side by side. By the very end of the 1960s, some black youth perceived Miller High faculty as being fair. Burt remembered: "Back then it seemed to me that everybody was treated fair and equal."

Still, although Burt did not "recall too many racial type fights in school, just kids disagreeing," and although in his remembered experience all students were "pretty much treated the same" by all teachers, black and white students mostly continued to stay apart.

> Back then the blacks stayed on one side, the whites stayed on the other side. The blacks stayed to themselves, and the whites stayed to themselves. Everyone once in a while, when we had to, we would intermingle. Other than that it was the blacks stayed to themselves and the whites stayed to themselves.[145]

Unlike Dotty, who remembered a more pervasive intermingling of students across racial boundaries, Nat and Burt recalled a continued separation between white and black students, a pattern that indicated perhaps a greater degree of racial border crossing between black and white women in general tracks than between black and white males in the same tracks. Annie's and Doris's recollections of the late 1950s, immediately following desegregation, hinted at timid communication across racial divides between white and black female students; Dotty's recollections described friendships and bonds established between black and white girls in the general track by the mid-1960s. One constant remained, however: For the general-track black and white students, both males and females, Miller High was not a place you visited after class hours.[146]

Thus the upper academic track, as earlier suggested and as revealed again in Burt's recollection, continued to be populated by those students with means, or at least with the familial support to engage in all aspects of school life, and the lower general track continued to be populated by economically disadvantaged students who could not fully engage in the life of the school.

By the time Burt Sadden graduated, a decade after the Halloween dance incident, dances were now safe for black students to attend. But by then, however, as Burt remembered, blacks chose not to go to dances.

> We could have [gone to dances] if we chose to. We didn't feel comfortable with the music that was being played at the time. It was a music thing.[147]

Thus, although African American alumni Nat and Burt, graduates in the 1960s, recalled a continued separation between white and black students, they did not explain the separation in the same ways that graduates of the late 1950s and the very early years of desegregation explained it. For 1950s graduates Annie and Doris, it was a "black and white" thing. Animosities between black and white students were strictly defined by skin color. For Dotty, Nat, and Burt, graduates of the late 1960s, the "black and white thing" acquired a more complex cultural dimension. Beyond the question of the skin color, students hung out with students with whom they felt most at home, and in the general track, black students felt most at home, culturally, with other black students.[148]

For Nat, upper middle-class African American alumnus, and for Burt, working-class African American alumnus, both of whom attended the general track, high school was something you had to do and couldn't wait to get out of, just as it had been for white alumnus Robert Heart who was of lower socioeconomic background. Unlike Robert Heart, however, Nat and Burt were not out to "beat the system" and prove they could stick it out. For Nat and Burt, finishing high school was a question of honoring their parents and fulfilling familial obligations. Burt explained:

> I did attend graduation, of course; it meant to me that I achieved what I originally set out to achieve as far as my family was concerned. This was one of the things my mother and father emphasized to me: you had to be at least a high school graduate to be partially successful in this world. Back then that was pretty much true. We couldn't go to college. It was more important to try and get a decent job and survive.[149]

Black males in the lower general track endured school, and just like their white male counterparts in the general track, they couldn't wait to get out. Nat recalled: "I did as little as possible. I wanted to get out and keep on going".[150]

Students of the Divided Generation of the 1950s and 1960s, as alumni recollections revealed, identified their places vis-à-vis one another and school authorities primarily along the socially inherited binary constructions of black *or* white, female *or* male, middle class *or* poor. Constructions of the middle-class category broadly included lower-/working-class status, by contrast to the very poor on one extreme and the very rich on the other, as we shall see in the next chapter. Recollections across categories of social identity also underscored school authorities' leniency toward girls in general, and the influences of the combined force of tracking and economic realities of college attendance in dividing students within a matrix of intersecting categories of gender, race, class, upper academic, commercial, and general-track identities and labels.

Thus, in the 1950s and 1960s, white middle-class females and those who could play the part felt most at home at Miller High, where they enjoyed a myriad of opportunities for self-expression. Poor white adolescent girls, on the other hand, shunned by peers and teachers alike, often dropped out or became invisible. Middle-class black female students, who felt least comfortable within what was to them a hostile and alien institution in the early days following desegregation, transformed their circumstances from marginalized students attending at best the commercial tracks, to college-bound integrated students attending academic tracks by the very end of the 1960s. While the very few middle-class black male adolescents who attended the academic track and excelled in sports felt empowered by their education at Miller High, young white males on the other hand, whether in the academic or general tracks, whether middle class or poor, and many black males in the general track, experienced high school as a place where one "did time," as an irrelevant and economically useless institution, although, as we have seen, for different reasons across categories of race and class.

However, while structural forces largely circumscribed the relational nature of Miller High students' experiences with each other and school authorities in the 1950s and 1960s, by limiting meeting places between richer and poorer, between whites and blacks, and by organizing poorer white and middle-class black females in commercial and general tracks, and poorer white and middle-class black males in the general track, Miller High students, through the lenses of their situated positions, and within the setting of a rural town in the 1950s and 1960s, gingerly negotiated those positions. Under the range of unrelenting hierarchies, some students found ways to cross boundaries of class, race, and gender, as I explore in the next chapter, where I also explore the meanings attributed by students to the social categories by which they identified their place and that of others at Miller High.

Chapter 2

Cautiously Negotiating Social Divides: A Conservative Student Body (1950–1969)

Within divided and dividing institutional spaces, some male and female students in the 1950s and 1960s practiced crossing class and race borders in their imaginations when they could not do so in reality. Overall, however, students' opportunities for developing friendships across gender, race, and class divides were restricted and shaped by institutional norms. Some students willingly embraced those norms; other students, by dropping out or doing time, perpetuated and reinforced those norms, making the Divided Generation the most conservative of the three generations.

Peer Relationships

Intersections of Class, Race, and Tracks

Class was a category of identity to which students of the Divided Generation were particularly sensitive, whether they were black or white. Overall, adolescents who came from wealth did not attend Miller High, and Miller High students were very much aware of that. The very wealthy youth attended the many well-known private schools in the surrounding areas. African American young women and men might know about them through their mothers, whose domestic work brought them in contact with the area's wealthy families. Some young white girls of lesser means might know them through short-lived courtships they imagined might change their family fortunes.[1] These wealthy students occupied virtual spaces in the minds of many who attended Miller High. Their absence at school was recognized only in contrast to their presence in the neighborhoods. Graduates across the 1950s and 1960s recalled the contrast:

> We knew that kids who went to private schools were different. And I can remember when we worked on the yearbook, we would try to go out and collect money...we were like these poor little village children knocking on the castle door...This was just where the rich people lived, and the rich people were different from us.[2]
>
> The richer kids, or the kids whose parents had money, wouldn't necessarily send them to public schools; they sent them to [narrator names all the private schools in the area].[3]

It was also their absence at Miller High that made the school middle to lower class.

> Everybody was middle or lower class. I mean, we didn't have any upper class. They all went to private schools. If they had any money, they went to private schools.[4]

Thus, attending the public high school of Miller Town meant attending a school less likely to be frequented by the very rich. Within the middle- to lower-class continuum of Miller High, however, students arranged themselves and others as poor or upper class. Within high school borders, a status comparison quickly replaced the area's virtually absent upper class with a high school "upper class" defined by Judy Law, Robert Heart, and those who considered themselves "poor" as those who were college bound, drove their fathers' cars, and spent their after-school hours involved in extracurricular activities. Thus Alice Web, who felt like the "village child" when knocking on rich families' doors for yearbook donations, constituted in the minds of students like Judy Law and Robert Heart the "upper class" of Miller High—those who could afford to participate in after-school activities, and whose fathers could afford to pay college tuition.

Still, while the area's rich lived behind what seemed like "castle doors," within Miller High, some students practiced crossing class boundaries. Young white women of lesser means who could look the part forged friendships with more privileged white women within clubs held during school hours. Thus Judy Law and Alice Web became friends through yearbook club collaborations. Black young men in the upper academic track forged friendships with white males in classrooms and on sports fields after school, as the stories of Norman Good, Willie James, and Jimmy Cole revealed. Poor white farm boys could cross class lines briefly during their senior year, when they might date a junior "upper-class" girl.[5] Sometimes, however, poor white boys just couldn't wait. Now and then, they'd force themselves into the "upper class."

> Us lower peons, we just kind of hung out with ourselves...they [upper class], because they were going from one academic class to the next academic class...would walk down the corridor and they talked to themselves...and some of us would maybe be invited to their parties. Not very often...well, if the upper class didn't invite, we would just go anyhow. Because we would find out. We would go anyhow and show up.[6]

Thus, for white young women, the requisite for class-crossing was "looking the part." For black young men, the requisite was excelling in sports and academically within the academic track.

For the poor farm boy, the requisite was grade-level status, and, sometimes, bully behavior. The very poor whites, those who lived in shacks, were not reported to cross class lines.

Middle- and lower-middle-class black young women were, until the late 1960s, mostly funneled into the commercial tracks along with white young women of lesser means, and did not participate in extracurricular clubs such as the yearbook and school newspaper—clubs overwhelmingly populated at the time by white young women of means. They might have found opportunities to interact with the more privileged white young women only during physical education, held during school hours, since they were also less likely to attend after-school activities. Testimonies suggested that the black young women of Miller High in the 1950s and 1960s, by virtue of being assigned to lower academic tracks, mostly interacted with white young women of lesser means.

Class distinctions, as already suggested, were inextricably linked to racial differences. Although Nat was considered "aristocracy" among his black peers, he continued to be considered "poor" among his white peers in the general and vocational tracks. Testimonies suggest that at Miller High in the 1950s and 1960s, regardless of evidence to the contrary, if a youth's skin was black, then that youth was poor in the minds of the poorer whites. Speaking of his economic circumstances, Nat Right recalled:

> Blacks were considered poor among whites. As far as we were concerned, we were the aristocratic blacks...Blacks thought we were so rich.
> *Your family was considered to be wealthy, but you came to school and the white kids considered you poor?*
> Right.[8]

The two different testimonies by African American alumnus Norman Good, who attended the upper academic track, and African American alumnus Nat Right, who attended the lower track, suggest greater prevalence in the lower tracks of intolerance among whites toward blacks. Norman recalled experiencing acceptance by richer whites in the academic track, while Nat recalled being rejected by white peers in the general track, even when his place of residence proved his economic standing.[9]

Norman's and Nat's contrasting testimonies bring to light poorer white students' denial of black students' possible economic success. The high status that the academic track held in the school to begin with, coupled with its demographic composition (overwhelmingly white females and males of means with varying levels of academic credentials, but only one or two black students with impeccable academic credentials), might have eliminated the pressures of race competition in the wealthier upper academic track. In the lower tracks, however, as testimonies suggest, perceptions of class differences were inextricably linked to the question of race. During the early years of desegregation at Miller High, as earlier shown, racial encounters between young black and white men in the lower academic tracks were tense and even violent.

It is important to underscore that a young black man entered the rarified air of the upper academic track only through persistent parental insistence, relentless self-discipline, and superior performance in academics and sports. While tracking funneled white students with means regardless of their academic merits into the upper academic track, a black youth with means would be systematically assigned to the general track. Thus a young black youth's class status was trumped by race. In the eyes of white administrators as well as white peers, particularly male peers of lesser means in the general or vocational tracks, a black youth was black first and only.

Male peers in the general track did not accept Nat Right's middle-class status. The general and vocational tracks were where some of the angriest and most disenfranchised young white men of the rural community faced the company of young black men, most of whom, while disenfranchised within the broader white community of their time, were by contrast fully integrated into their black community and steeped in family values that emphasized respect for parental authority and familial obligations. Resistance to Nat Right's middle-class status by white peers in the lower tracks might suggest that students pejoratively called "hillbillies" by middle-class white peers felt doubly threatened by black youth who held many of the same values—and the same Christian affiliations—as middle-class white peers.[10] Denying black young men middle-class status would have negated the possibility of their social superiority in a world where the poor white youth were considered the least for having the least.

Intersections of Gender and Race

Whites and blacks more easily crossed racial boundaries within same-gender relations. However, crossing racial and gender lines simultaneously was problematic. Although encounters between young white women and young black men began emerging in the 1960s, as Dotty's recollection suggested, interracial coupling at Miller High was taboo throughout the 1950s and 1960s. Black and white students who might have been attracted to each other kept their desires secret. Doris Right, a black female alumna, recalled a white male classmate confiding to her years later: "I used to watch you in class. I really liked you, but you know, we couldn't say anything." Doris also confided her own repressed interests: "You know, you would look at a person [white male] and say, 'Oh, gee, he's cute,' and that would be it. You wouldn't even entertain the idea, you know, of any type of relationship." By the mid-1960s, one rare student couple ventured crossing racial and gender borders at Miller High. Dotty recalled that the young man was black and the young woman was white. Still, the pair's dating was more private than public, and the couple stayed away from school socials.

It is perhaps no wonder, given the social taboos surrounding interracial couplings, that racial encounters would be most violent at high school social events that involved dates. African American alumni recollections of the Halloween dance vividly depicted the tensions involved.[11] When it came to the dance floor, social status quo of the time prevailed. White racist students, buttressed by their white teachers' tacit support and significantly outnumbering black students,[12] claimed full control over the dance floor.[13]

Social status quo further prevailed in the power differentials manifested in dating choices. White males of means exercised the greatest choice in terms of whom they could date among white young women, since they could date white women of lesser means, as Judy Law's testimony showed. Next to them, white young women of means exercised the most choice by actually rejecting poorer boys, and accepting their advances only if the young men were seniors and they themselves were of lower grade level, as Robert Heart's recollections suggested. Black males and females exercised the least choice when it came to dating within their high school community because there were too few blacks to form a big enough social network for dating.

Beyond dating and the dance floor, power differentials in gender relations within everyday high school life were different for white and black students. White girls, graduates' stories have suggested, patronized white boys, whether by doing their homework or taking charge of all social activities related to the school, and the boys allowed them as much. In relating with one another at Miller High, white males and females reflected behavior more akin to the private realm, where, during the 1950s, as historians have noted, "women could be a solution to men's dilemma... [and] provide men a haven in a heartless world,"[14] suggesting a kind of domestic takeover by white middle-class female students at Miller High. White boys let them have the run of things, not unlike white men of the time who let their wives have the run of all things domestic, including volunteer work.

As historian Linda Eisenman suggests, "The 1950s encouraged women's activism through... civic minded organizations such as the League of Women Voters... Young Women's Christian Association... Women could participate without committing themselves to regular, paid employment."[15] While Miller High's white boys lived on the edge of classroom participation, working before and after school, sprinting out of high school windows to join firefighters at a moment's notice, or taking leave to go hunting, white female students were stable residents. When they did leave school property, it was to go on errands to shop with money already provided them by their elderly teachers. A short leap of logic explains the seeming disjuncture between young women holding leadership positions in the Future Business Leaders of America and on the editorial staff of the school newspaper without corresponding leadership positions in the world outside high school boundaries. As long as the work remained unpaid, white young women were encouraged to develop their leadership roles.

This also helps explain why white boys of lesser means who were not college bound placed little value on attending classes for which one received no pay. For them, the "real" work occurred outside of high school in the "real" world. So much so that some of them risked, in a gesture of ultimate disempowerment of school authority, punching out the demanding teacher and forsaking school altogether for the possibility of fortunes made in the "real" world. Because college, which held promises of money and ascendancy into upper-middle class, required money to attend, being a student did not pay for poorer white boys. As for "richer" white boys, if not college, then jobs in their fathers' businesses often awaited them, suggesting a continuation into the 1950s and 1960s of a pattern of nonchalant attendance among white male students, which historians of education have identified with the beginning of the high school in the early nineteenth century. Regardless of whether they graduated from high school, these young men were guaranteed jobs.[16] In general, college-bound white girls and boys prepared for "middle-class families [that] came to dominate the era...a white-collar husband, supportive wife, and several children residing in a comfortable home."[17] Judy Law explained how "you were meant to grow up, marry the boy next door, go to the local church, and stay in the family."

By contrast to the lives of middle-class white young women, middle- and lower-middle-class African American young women headed straight home after school. During the early years of desegregation, they might have sought the help of their more popular male counterparts to stop some white boys' misbehavior toward them. However, white boys' more offensive behavior toward black girls were furtive, and the culprits went unrecognized as they anonymously and perversely "felt behinds" in the hallways. For middle-class young black women of the early days of desegregation, school was not a safe place. Whether sexually offensive[18] or less sexual and more annoying, as in DiPaglia's teasing of Annie, some white boys' behavior toward some black girls at Miller were experienced by black girls as aggressive and sexual. But it wasn't just white boys' aggressions they dealt with. Some black girls fought with black boys and black girls because they looked different. Doris Right remembered:

> Some of the black kids didn't like us... We were different, not that we knew it, we didn't know that we were any different that any other black kids. My sister and I have the green eyes—nobody during that time had green eyes—and they would call us "gray-eyed." Whatever, they weren't even gray, they were green, and they'd fight us because the color of our eyes... And one guy, he was a big guy... he would want to fight me every day.[19]

Doris's description of her relationship with other blacks echoed historian Franklin Frazier's descriptions of black students in the earlier part of the twentieth century, when "discriminatory practices were visited by light-skinned students on darker ones."[20] However, in the case of Doris, something else might have been at play that explained the discrimination she experienced by other black youth, since Doris's black skin is darker rather than lighter. Doris and her sister, although light-eyed, were also the daughters of parents who owned land in the more prosperous part of the town. Middle-class black female students in the early years of desegregation at Miller High experienced their bodies as targets for white boys' furtive stabs at sexual misconduct or for other blacks' assaults. That their persons, their emotional and rational selves, were submerged under the combined weight of their gender and skin color in the early days of desegregation was made even more apparent in contrast to the lack of reported aggression toward white women. Immediately following desegregation, racial inequities were compounded by gender inequities for Miller High's young black women. By the time Dotty attended high school in the mid-1960s, as earlier discussed, life at Miller High for young black women felt more comfortable.

Intersections of Race and Extracurricular Spaces

Against resistance by many whites who never ate with blacks at lunchtime, white and black students would occasionally sit together at lunch. Either the black student was a star athlete who sat with white star athletes, or pockets of white students were neighborhood friends of some black students and had lunch with their childhood playmates. Speaking of her white friends, Annie Cole recalled:

> You would have a few that would intermingle with you [in the cafeteria]. Like the kids that we played with in the neighborhood would come over and sit with us. Like Ike and Dewey and Charlie, and all of them...Some of it was nice. We weren't all bad.[21]

Knowing your neighbor made sitting with her at lunch a natural extension of your life in the community. Although blacks were still not allowed in restaurants reserved for whites during the early years of school desegregation, some Miller High black and white students were eating together at the same table in the public setting of the high school cafeteria. There they could practice new social mores within the protection of the law. However, young blacks and whites overwhelmingly segregated during lunchtime, as underscored by African American alumnus Nat, who attended the general track, and white alumnus Bud, who was college bound. Thus the cafeteria, the one place within the high school that housed the greatest numbers of male and female students of different racial and economic backgrounds in one place at one time, never fully became a place of integration, even as some students ventured to cross identity boundaries. On the contrary, it would become, as we shall see in later years, the barometer for levels, not of integration, but of segregation (except for the Border-Crossing Generation), and a place more akin to a prison mess hall than a young people's lunch area.

It was on the sports field after school that black and white males practiced becoming friends. Black and white females experienced friendly relationships on sports fields in physical education during school hours.[22] The one space within Miller High where students most consistently crossed racial borders (but never gender borders) throughout the 1950s and 1960s was in team sports. However, access to team sports after school required student availability, which poorer students didn't have, and which middle-class black young women did not have during the early days of desegregation, because their parents requested their presence at home immediately after school for reasons of safety.

Implications of School-Shaped Associations for Cross-Racial and Cross-Class Encounters in the Workforce

While relationships across gender proved more empowering for white middle-class young women and less empowering for black young women, relationships within gender proved to be more conciliatory among white and black women in the commercial and general tracks than among white and black males in the general tracks. Non-college-bound young women, black and white, were more likely to be enrolled in the commercial track. In view of the fact that preferential treatment was given to girls, as reported by white and black male and female graduates, it is not surprising that in general young women might have had an easier time associating with each other across racial boundaries within lower academic tracks.

However, while the tracking system funneled poorer young white men and black young men regardless of economic background into lower curricular tracks that prepared them for menial labor, it funneled young white women of lesser means and black young women across economic backgrounds into lower curricular tracks that prepared them for secretarial white-collar jobs. This gender division across race in lower tracks opens up for investigation the implications of gendered differences in tracking in the 1950s and 1960s (females {black and white} = commercial; males {black and white} = vocational) for relationships continued into the workforce—privileging perhaps white and black women's opportunities for crossing class lines over white and black men's opportunities to do the same. Secretarial work generally brings one in contact with people of different economic backgrounds, while menial labor work mostly keeps one isolated from daily contact with people of upper-class backgrounds. Given the flourishing small black-owned businesses of Miller Town in the 1950s and 1960s, there would have been opportunities for young black women to find employment as secretaries and to encounter people across both class and racial lines. Further research is needed to investigate whether all women have greater opportunities than men to establish relationships across class and race lines in the workplace at large.

Youth and Authority

Constructing Teacher-Student Relationships across Social Categories

Overwhelmingly, this was a teacher-centered generation. Except for college-bound white alumnus Bud Land, who graduated in 1968, all alumni interviewed recalled particular teachers who helped them and who eased the difficulties of high school life: the gym teacher for lower-middle-class African American alumna Annie Cole; the shop teacher for poor white alumnus Robert Heart; the older female teachers for poor white alumna Judy Law; the coach and journalism teachers for college-bound white middle-class Linda Moss; and so on. In some graduates' reports, favorite teachers taught subjects that students loved and excelled in;[23] in other reports, favorite teachers were just or inspiring, even when students neither loved nor excelled in the subject matter. Sometimes the teachers were close to the students in social status, as in the case of the shop teacher and Robert Heart; or close emotionally and in world views, as in the case of Annie Cole and the gym teacher. Annie Cole loved her gym teacher not only because she appreciated Annie's athletic skills, but also because she treated blacks and whites equally and was someone in whom Annie could confide. Judy Law felt sustained by her elderly white teachers because in their company she was appreciated for the middle-class things Judy longed for and strived to live up to, things she was denied in her isolated and poor country life at home: looking pretty, writing poetry, winning awards. Linda Moss loved the teachers who believed in her academic abilities, and so on. These hero-teachers often filled voids in students' lives, helped them to see worlds beyond their own, and to gain freedom in imagination if not in reality. In the 1950s and 1960s, in a mostly rural small town, with a very small high school population and an economic and social climate conducive to class mobility, the individual teacher could become the agent of change that interposed motivation, inspiration, vision, and fairness between a student's social background and the comprehensive high school's often-documented bureaucratic tendencies for social reproduction.[24]

However, good relationships with teachers did not necessarily translate into a good education for all students. The mosaic of recollections from white graduates of the 1950s and 1960s sketched a high school education that either failed to prepare them for life after high school or was irrelevant to life outside of high school. College-bound graduates were surprised and shocked at the level and quality of work required at the college level. They felt completely unprepared for higher education, and graduated from Miller High with few study skills and a weak academic foundation.[25] By contrast, African American alumnus Norman Good, who attended the academic track, remembered receiving a solid education: "I feel as though I got an excellent education at Miller." Recollections, as earlier discussed, also revealed that among the college-bound white students, excelling in academics was not as crucial as it might have been for a black student who would have had to invest every ounce of effort to prove that he was worthy of the academic track. Accordingly, his efforts would have paid off in academic achievement and scholarly preparedness. Thus the academic track for white students was less about academic aptitudes and primarily about economics, particularly for males.

To both college-bound and non-college-bound white males, academics seemed irrelevant to "real life" as they knew it in Miller Town. White males of means would, as documented by historians for generations of high school students before them, inherit their fathers' businesses, or "become clerks, tellers, salesmen, or agents, and were thus able to maintain their socio-economic status without the effort of long term study."[26] However, although by the 1950s young white men could drop out without much economic repercussion, they could no longer drop out voluntarily without negative social repercussions.[27] Thus, many "did time." Their "real lives" were defined by their fathers' businesses. White males of lesser means also did time as they continually juggled jobs on the side. They too perceived their academic education to be irrelevant and disconnected from the realities of their lives. These white male students, richer and poorer, dissatisfied with the "system," turned out to be perhaps Miller High's most conservative force. Not only did they "suffer" the academic system, sometimes out of loyalty to their parents, and relinquish the school's social life to their female counterparts, but also inadvertently supported, by their complaints of the irrelevance of an academic curriculum, the public high school status quo since the triumph of the comprehensive high school.[28]

During the 1950s and 1960s at Miller High, high school education would prove most valuable to non-college-bound white and black women and to the very few black males who attended the academic tracks. For these students, the high school diploma meant possibilities for social mobility and expanded terrains of freedom. These students also proved to be a conservative force in the perpetuation of the comprehensive high school status quo, because they invested their energies in using its education within its prescribed modus operandi to better their lives.

A Conservative Student Body across Categories of Social Difference

On the whole, students at Miller High acted much more conservatively than sociological case studies of 1950s and early 1960s high school life suggested. Coleman, the education ethnographer most notable for identifying high school youth of the 1950s as focused solely on popularity status among peers, painted them as primarily engaged in their own adolescent societies, where "looking good" determined hierarchies of popularity. He underscored how little influence adults held in the lives of high school adolescents.[29] This work, on the contrary, reveals young white females' negotiations of appearance as a means to maintain closeness with teachers and school authorities who more readily accepted a certain look in their charges. As alumni reports revealed, young white women's behavior at Miller High was not as much about popularity among peers as about maintaining good relationships with teachers and practicing being pretty and successful. Acceptance by authority figures whom they considered role models was of utmost importance to them. These white young women were a very conservative force, as Coleman's study suggested, but at Miller High they strove to please authority figures with whom they shared almost familial relations. Their focus was on ensuring their particular *student* status, rather than *peer* status.

Coleman's assertion that "a working class boy or girl will be most left out in an upper middle class school"[30] does hold true in this work, evidenced in hierarchies of dating patterns and Robert Heart's acute awareness of class distinctions. However, at Miller High, as alumni testimonies suggested, students—female and male, black and white—befriended or shunned each other within the predetermined tracking structures, suggesting that the school's organization of students within predetermined groupings influenced to a much greater degree their associations with each other than their own volitions. In fact, as recollections of white and black alumni revealed in this work, many students wished to associate with each other across dimensions of class, gender, and race, but, as Alice Web expressed, "We were to a certain extent in our slots."

On the whole, students of the Divided Generation participated in perpetuating the high school life of the early twentieth century, where "commitments to competition, conformity and individual merit" prevailed,[31] by the traditions they embraced from previous generations, and by what they didn't contest. The "doers," as Linda Moss called them, made sure to muster school spirit, to organize Miller High's social life, and to please their elder role models. Black males in the academic track relished the opportunity to compete with their white counterparts and to prove themselves. Poorer students and those who experienced school as an unfriendly if not outright hostile place dropped out, leaving the run of things to the "doers." Students for whom school was irrelevant or hostile, but who stayed, "did their time." Although the story of Miller High students' lives during the 1950s and 1960s is one of many intertwined stories, on the whole the school's transmitted hierarchical divisions heavily organized their lives.

While a fair and benevolent principal helped the small student population navigate the highly charged period of desegregation with less rather than more damage, and while a very few teachers helped build hope and self-esteem in students across social categories, students lived lives mostly segregated along gender, race, and class divides. Entering the world of Miller High in the 1950s and 1960s meant stepping into the habits and traditions transmitted and upheld by white middle-class female students and teachers. It was a high school ill-equipped to help students relate with each other across categories of identity, even though it offered greater possibilities for democratic engagements among students across racial, class, and gender divides than the rural town it served. It was where students by and large remained close to their own across categories of identity.

School structure, rather than youth culture, overwhelmingly determined how Miller High students understood their place within school and how they imagined their place outside of school in the 1950s and 1960s.[32] Overwhelmingly, alumni recollections revealed that tracking divided students by class, gender, and race. Tracking had divided students by class and gender before desegregation in 1956, and then by race following desegregation and throughout the 1960s.

Carnoy and Levin's suggestion that high school has consistently provided a "social experience that is markedly more egalitarian" than anything the workforce might provide does not hold in the case of many students of Miller High in the 1950s and 1960s. This was not the case for white male students of privilege, whose work opportunities were guaranteed regardless of success in school; nor was it the case for poorer white males or middle-class African American young men who attended the lower tracks. For poorer youth, as recounted by those interviewed, school was a place of shame and took time away from earning sorely needed money. For the black male youth attending the lower tracks, Miller High became a place to suffer the full-blown force of racism. While a lone interracial couple would emerge at the end of the 1960s, and a very few students crossed race divides in an otherwise overwhelmingly segregated cafeteria, students by and large segregated by gender, class, and race.

Carnoy and Levin's observation does not however apply to white middle-class girls who practiced roles denied to them in the workforce. Overall, Miller High in the 1950s and 1960s most consistently and effectively privileged white middle-class adolescent women more than any other student constituency, since they would prove best suited to play roles agreeable to white authority figures, and to reproduce within and for Miller High the dominance of a white, conservative, middle-class lifestyle.

But by the end of the 1960s in Miller Town, with the introduction of football and marijuana, came the end of an era of formal manners and clear hierarchical relationships. Beginning with the 1970s, *in loco parentis* would no longer guide teacher-student relationships. As one alumna, a 1969 graduate, remembered: "We were the last class that wore skirts. The class after us, the class of 1970, really changed a lot of things."[33]

Part II

The Border-Crossing Generation (1970–1985)

Chapter 3

Memories of Interracial Peer-Group Affiliations: Integration Years (1970–1985)

In the 1970s and early 1980s, Miller High students began their school year as generations before them had: with the ringing of the bell by the faculty member with the most tenure. However, while they also inherited the *Key* yearbook, the *Miller Chronicle*, and the proms, the students looked and behaved in ways that often shocked the aging and retiring teachers. Young women could now wear pants, now and then some students might "streak—take all their clothes off and run through crowds,"[1] some young men wore their hair long, and some African American young men and women wore theirs in Afro style.

By 1972, sixteen graduating classes had attended desegregated schools. Moreover, while Miller High elementary feeder schools were mostly populated by white students and only handfuls of black students, some white children grew up as minority students in the one predominantly black elementary school in the mostly black neighborhood. As one white alumnus recalled:

> My friends in first grade were black. I went to the black school. After everything was integrated...they put the overflow [white] kids in the annex on Brand Avenue...They [the black kids] were my first friends. When I see those girls, Maggie and Loretta—I saw Loretta two weeks ago—it was just the greatest thing in the world![2]

While still known for its predominantly rural flavor, throughout the 1970s and early 1980s Miller Town was developing into a suburban settlement with a variety of housing types, from single-family homes along sidewalked roads perpendicular to Main Street to ranchers in gated communities for business executives. It was also in the early 1970s that black families' houses on Brand Avenue finally received indoor plumbing. The "shacks" in which the very poor lived, as reported by 1950s graduates, no longer existed in Miller Town by the 1970s; the poorer whites lived in rented apartments, often above shops or taverns along Main Street. Miller Town's population became increasingly white and suburban during this period, with the intensification of white flight from city limits, and "the local people, both black and white,"[3] people whose families had lived in Miller Town for several generations, began uniting against newcomers. This bonding of "old timers" would continue to intensify and reach, by the 1990s, a defiant tone, marking insiders from outsiders.

Within the walls of Miller High, the Future Farmers of America club was no longer offered; home economics was renamed bachelor living and offered equally to boys and girls; Future Nurses of America was renamed Health Careers Club; and sex education and drug prevention classes were added at the end of the 1970s. It is also during this period that younger teachers were hired to replace the retiring "old timers." This change created a division between young and old teachers, as young teachers demanded more academically but less behaviorally, and older teachers demanded more behaviorally and much less academically. Furthermore, the infusion of younger teachers, as alumni testimonies suggested, participated in destabilizing power relations between teachers and students: older teachers were less respected, and younger teachers sometimes crossed borders of familiarity to the point of having affairs with students.

During this period, nestled between the early years of desegregation and the immigrant wave of the late 1980s and the 1990s that flooded Miller Town with foreigners and city escapees, Miller High's racial composition fluctuated. With the intensified white flight in the early 1970s, the African American population decreased to two percent of the total Miller Town population,[4] compared to eight percent in 1950. By 1980, however, the relative population of black residents substantially increased to thirteen percent of the total population of Miller Town as white flight stabilized. Other things also changed—the educational backgrounds of Miller High attendees during that period were qualitatively different from those of the previous generation. These black and white students attended school together since the elementary level and doubled the size of Miller High graduating classes. By the 1970s, while many schools across the country were only beginning to desegregate, Miller High students had grown up in desegregated feeder schools. They knew each other not only as neighbors whose lives intersected on the streets and in the market place but also as classmates who had forged interracial relationships.

This generation further differentiated itself from the previous one in that alumni remembered one another less in terms of class and race and more in terms of peer groups consisting of jocks, preps, eggheads, potheads, motorheads, fireheads, musicians, and so on. This period also began with the introduction of football in 1969, which had been banned by the school throughout most of the twentieth century, as it was considered too dangerous for young men.[5]

The introduction of football, in turn, brought to prominence male football players, who, during the 1970s and early 1980s at Miller High, were particularly favored and remembered specifically as jocks: "The jocks were the group who played football."[6] The absence of football throughout the 1950s and 1960s most likely explains the rare use of the term *jock* by alumni who graduated before 1970.[7] All interviewed alumni from this in-between period that defines the Border-Crossing Generation recalled male jocks, black and white, as holding high-profile positions in the daily life of high school. To understand Miller High students' differently situated experiences during this period, one must first understand the high school world that black and white, male and female jocks participated in creating, for it is against this backdrop that the potheads contrasted, and the orbit around which the musicians, eggheads, and others revolved. As we shall see, it is also at the very periphery of all peer-generated groups that students who did not belong to a group continued to segregate by race and class.

Interracial Male Jocks

African American David Randle and white alumnus Josh White, both graduates of 1976, and Jim Garnes and Tim Whittle, white graduates of 1978 and 1981, remembered their high school careers not as students, but as jocks.[8] For these young men, the first day of school was an easy transition from middle school. Having trained together throughout the latter part of the summer, they were at home with one another before classes even started:

> We played football, so we knew each other. We got there in August. In July we'd do a weightlifting thing, start practice by the time school started...we were already acclimated. I knew what my classes are. You get your schedule ahead of time. It was very easy to start school.[9]

As testimonies revealed, a jock, whether white or black, richer or poorer, sat at the jocks' table in the cafeteria. When non-jock alumni reflected on their jock peers, they remembered them as a group unto themselves, those whom other males in particular perceived as having more fun. Musician Michael Hallner, graduate of 1976, recalled wistfully: "The jocks didn't hold back, they had fun...dated the best-looking girls in school." An indirect peek at Ronnie, a black jock graduate of 1974, through Josh's recollections, revealed a black jock fully aware of his hierarchical place, rendered by his jock status as well as his formidable physical strength.

> I remember one time I was walking through the halls in high school, and this black guy, *big* [alumnus's emphasis] black guy, played football, he was standing in the middle of the hallway, and he was intimidating everybody. He wouldn't let any one walk by. He was just playing around. I knew him, I knew he knew my dad [sports coach] pretty good, so I walked up to him and said, "Hi, Ronnie, how you doin?" He said, "I'm not touching you, your old man will kick my butt."
> *Were people intimidated by Ronnie standing there?*
> Yeah. He was big and all, but he was kind of a teddy bear too. Not that I would tell him that.[10]

Alumni who had been jocks admitted enjoying a highly privileged position in school:

> It was a huge social stature to be on a football team and being a good football player. I had one class I didn't pass anything, it was chemistry, I got a C all year and didn't do anything. I know that's why I got it, so I could keep playing football...Athletes back then did get away with things.
> *Did you take advantage of that?*
> Heck, yeah! We got to use the private bathrooms where the other kids couldn't. Coach let me have the keys.[11]

While David, Tim, Jim, and Josh shared memories of fun times as jocks, their recollections of non-jock peers differed, particularly for Tim, a white jock, and David, a black jock. Tim's recollections revealed a sense of righteous responsibility, and David's revealed a sense of systemic injustice and regret.

Tim, a 1981 graduate, remembered himself and his friends as protectors of people he considered weaker, among them the special education students and the girls. Tim's recollections suggested a self-appointed romanticized role not unlike a white knight—a righteous jock righting wrongs, fighting off the bad jocks:

> "You had a few kids that'd run their mouths always, assholes of the school. Sad to say, most of the time they were athletes, which was embarrassing for me. I got into a fight with a couple of them. Because they were teasing kids. I'd never do that. My friends and I never did that. You don't go fight the mentally challenged. That was taboo. Hands off. I got into a fight with some guy because he was teasing a girl, the way she walked. I said, "Leave her alone." He pushed me. I punched him.

These jocks, as Tim depicted, were also the protectors of blacks against white racists. Tim remembered:

> Me and Jack and Mat [came across] this guy [who] said, "We don't want them n— in here." [Me and Jack] beat the shit out of him, we beat him up in the parking lot, and Mat (the black friend) didn't even touch him.[12]

African American alumnus David Randle, who graduated five years before Tim, corroborated Tim's testimony, suggesting that Tim and his friends did not act in isolation. David recalled:

> [Racial slurs] were kind of taboo in that day. They would get on each other—I'm talking about white kids. They didn't even use the word. You know, it was a problem for them.[13]

In contrast to Tim's memories, however, David's recollections of other black non-jock peers were filled with regret. While Tim walked around "beating the shit out" of bad guys, rescuing those he considered weaker, and protecting the honor of his black friends, David spent much of his emotional energy explaining to his black peers why he played ball. David recalled:

> The people my age they were, I don't want to say militant, but more active into their culture. They would ask me, "David, why you playing ball for them, man?" I'd say, "'Cause *I want* [David's emphasis] to play ball." "They're just using you," they'd say.[14]

David's testimony revealed a frustration with the white and black faculty who did not do more to actively recruit capable black young men into sports; young men who, as he saw it, needed to be sought out rather than left on their own to join a system they understood to be working against their best interests. David's recollections emphasized the importance of one's involvement in an integrated team sport as a means not only of self-expression, but also as a means for overcoming racial divides.

> It doesn't seem right to say that some of my best friends are white, don't sound right, but I don't know a better way of saying it. See, sports is different. You do bond there. You don't lose that. You gotta trust one another. Race doesn't play a factor into that when you get into a football team.[15]

David's love of sports competed with his loyalty to black peers. However, it was also his privileged position as jock in general, and the bonds that it allowed him to develop with white peers, that created the opportunities for David to introduce his black peers into the world of middle- and upper-middle-class whites:

> They [the rich white kids] had nice parties...I got in because I played on the team...I used to go by myself to the parties, but by senior year I used to bring other black people.
> *Was that a good thing?*
> That was a good thing. That was a good thing...it brought everybody closer. Then you go back to school Monday morning, you just spent all Friday night, all Saturday night together, it would bring everybody closer together.[16]

By his senior year, David had begun using his unique position as a black jock to actively create opportunities for black peers to cross racial divides. Not unlike Tim, who somehow felt responsible for the well being of those unjustly bullied or discriminated against, David felt responsible for his less privileged black male peers. None of the jocks interviewed reported thinking much about life after high school, and all of them attended school willingly, even happily. However, how and why they made it through school differed for David from his white peers' experiences.

For David, going to school meant going against the current of truancy among his non-jock black peers. He explained:

> I loved high school. I went to school every day. Some of my friends, they'd cut school all the time. [After some time] they knew I was going to go to school. They wouldn't even say, "David, are you coming with us?"[17]

David's genuine enjoyment of high school might not have been idiosyncratic. Other graduates' recollections suggested a shared playfulness and ease among black jocks at Miller High during the 1970s. Resisting peers' truancies seemed easy for David. It was outweighed by his passion for football and acceptance by white peers.[18] David explained his non-jock black peers' alienation as the result of an unresponsive school system coupled with an upbringing that did not discourage using race as an excuse for not participating in school:

> I think those kids that the system beat them...for one reason or another the teachers kind of brushed off [the black kids]...A lot of black kids didn't participate in extracurricular activities. Nobody really approached us "do you want to join this or that?
> I made it 'cause of my mother. "Don't come home tell me the teacher did this or did that"...the teacher was right no matter what happened. We were never allowed to use race as an excuse...If you want to do it, do it.[19]

David's willing attendance at high school, however, was solely and primarily linked to his playing football. His recollections of Miller High teachers were critical.

> For me it wasn't a very positive relationship with teachers. They kind of brushed us [the black kids] off a little bit, and we were just like there...I never met a guidance counselor in high school. I mean, they had my grades, they knew I wasn't stupid, but they never mentioned college or anything.[20]

Not only did he remember feeling frustrated with the mostly white faculty who ignored black students, but also with what he called the "wrong attitude" of the black male teacher who taught black history.

> He had a good old boy attitude...He even told us, "They're making me teach this class for obvious reasons."...But he didn't really get into black history. He could have been a lot stronger.[21]

David's recollections of Miller High faculty were far more critical than those of black graduates in the Divided Generation, suggesting a greater sense of rights to better services on the part of black graduates of the Border-Crossing Generation than previous black generations of Miller High students. David's insistence that faculty failed to reach out to black students resonates with the late-1950s poorer white males' demands for sensitivity to students' needs, a theme also echoed, as we shall see, by Tim Whittle, a working-class non-college-bound white jock. While David enjoyed a privileged status among peers and with the coaches, he continued to feel marginalized as a black student. His words, "I mean, they had my grades, they knew I wasn't stupid, but they never mentioned college or anything," echoed Norman Good's disappointment with not having been given options by guidance counselors to go to colleges other than the historically all-black colleges. The difference between David, a 1976 graduate, and Norman, a 1959 graduate, was that David expressed exasperation while Norman expressed sadness. Although frustrated with Miller High teachers, there was never any question in David's mind that he would not graduate. As many graduates across the two generations reported, honoring parents' wishes was a big part of graduating: "I never thought about not graduating... I think it was a big deal for the parents." Although David knew that he had to graduate, he never thought about what he would do once he did graduate. It was David's coach, Luke Dare, who suggested to David that he consider college and who helped him to apply:

> Luke Dare, my football coach, he just said, 'David, what are you going to do with yourself?' He said, "You going to college?" I said I hadn't thought about it. He said, "You need to think about it." And he set it up. He called the school and got the [football] recruiters to come over, and he took me over to the college and he introduced me to the football coach... I probably would have gone down one of those factories in Miller Town, but I went to college.[22]

David's status as jock and high-performing ball player opened doors into a world he hadn't previously considered. Although his life path turned out much better than he had envisioned, David lamented that many of his black peers whom he considered more deserving than himself did not participate in sports, where they could have shined and, as he remembered, could have enriched Miller High. David explained:

> There were good athletes from my neighborhood, but nobody really approached us about do you want to join..., athletes that were better than me... The school lost a lot by not getting more of the black kids involved.

David's recollection of his lesser athletic abilities but full extracurricular involvement compared to the superior abilities of his black peers on the one hand, but their lack of school participation on the other, underscored the precarious nature of high school success for black young men. As David understood it, it was a success that required, beyond mere attendance, an immersion in sports life both before and after school. Moreover, it required that the student actively seek participation in the absence of school authorities' invitations to do so. Thus the onus for success at Miller High lay squarely and solely on the shoulders of the black male student. Were black students determined to seek inclusion, they would not be refused, but neither would they be recruited. David recalled:

> Nobody approached us about... so what happened... a lot of kids, they didn't get involved in anything extra. They just wanted to get out of there.

2:15, time to go, we're out of here... It wasn't fun for them to go to school. I can tell you that. That's why they didn't go.

For David, a high school education became a passport out of factory life into higher education through a football scholarship that changed the course of his life, but for Tim, Jim, and Josh, it became a way to avoid hard choices and ultimately forfeit an academic education, choices that were facilitated by a laissez-faire older faculty and lax graduation requirements. Echoing the stories of white male graduates of the Divided Generation, white jocks of the Border-Crossing Generation continued white young men's nonchalant attitude vis-à-vis studies and reliance on their white female counterparts to do their daily homework. Tim recalled:

> I didn't have to work hard. Girls did my homework. I never read a book in high school. I'm embarrassed, kind of, I still don't like to read... To sit and read a book, I sooner chop my finger off. It is the most boring thing in the world for me. I never read a book. I never did homework. I didn't have to do it... So, far as I was concerned, my education was poor.[23]

Their status as jocks accentuated the ease with which they got through the system and with hardly any effort invested in learning. In large part it was the structure of the comprehensive high school that made it easy for them to buy free time as students by accumulating certain amounts of credit points.[24]

Miller High white male jocks of the Border-Crossing Generation, richer or poorer, knew they would find jobs after high school whether they attended college or not. Tim, who was the poorer jock and whose parents could not afford college, went to work for a well-established company in the area where he climbed the salary ladder. Jim Garnes went to work for his father and eventually took over the family's small business. High school education for the young white male jock of the 1970s and early 1980s, as it had been for white young males in general at Miller High at least since the 1950s,[25] was not about academic learning. Just as teachers in the 1950s looked the other way during hunting season, or when boys climbed out of windows to run to the fire station, in the 1970s, many of the older teachers' lax academic expectations conspired to lessen the importance and value of academic learning in the minds of their white male students.[26]

Jock Jeremy Garnes remembered:

> At high school level [the courses] were not useful and I could have done without high school... I remember taking a physics course, it was literally a joke... we would play the whole time... came test time, we never

> studied. He'd give us a test and walk out the room. He had copies of the test out there. One or two people would go out there and tell us the answers. Back then the older teachers just wanted to...get done.[27]

Teachers who were older, according to testimonies, gave easy grades, but also were more likely to send their charges to the principal's office for misbehavior. Younger teachers challenged their students academically but were more lax with discipline.

Lax academic standards on the part of the older teachers contributed to Miller High's own version of what Labaree identified as credential inflation, and a reciprocal laissez-faire academic attitude on the part of jocks whose energies were already focused on sports rather than studies.[28] Younger teachers' more demanding academic standards, on the other hand, challenged the jocks' supremacy. Tim's recollections suggested that certain young female teachers in particular didn't appreciate the jocks:

> I had this trigonometry teacher I hated, she was young...I remember one day she told me, "You're going to be a loser." So I was—hold on a minute, so if I don't pass this class I'm going to be a loser? I won't grow up to be successful—I did this in front of the whole class—I won't grow up and be a successful math teacher like you and make thirty thousand dollars a year? From that point on we hated each other.

Tim's arrogant reaction to the teacher's bruising indictment of his abilities was not unlike how Robert Heart and his friends reacted to teachers they considered offensive. Furthermore, Tim shared a lower socioeconomic status with his predecessor of the Divided Generation, Robert Heart. Although not as poor as Robert Heart reported to have been, Tim had not counted himself among the college bound. These alumni recollections suggest that less affluent, more aggressive white males, across generations of Miller High graduates, continued to publicly assault their teachers verbally or physically. White male narrators across time also continued to justify their aggression toward school authorities by underscoring teachers' unjust or insensitive behaviors. Tim recalled how a peer whose parents had just divorced reacted to a teacher who, according to Tim, "was an idiot":

> I remember Andy Dobbs's parents split up. He was very emotionally distressed in tenth grade. The first day back to school she told him, "Just because your parents are split up don't mean you're going to get away with anything." He started shaking. He turned five shades of red, picked up the desk, and threw it at her. Almost hit her. She fell on the ground and I started laughing.

Tim's memories also echoed those of 1950s graduate Robert Heart in that Tim and his friends would step up to the academic challenge only if the relationship with the teacher was a good one. Tim recalled a particularly helpful teacher:

> She made you learn, she was actually a very good teacher. The age of the teacher was actually a big deal for me, and athletes back then did get away with things... [but younger teachers] they wanted you to learn. If they had to put up with you being a bit playful or something, as long as it did not get out of hand, sometimes they would join in.

Just like Robert Heart and his friends of the Divided Generation, Tim and his friends of the Border-Crossing Generation attended Miller High not as recipients of instruction but as critical consumers of its deliverance. For them, teachers were not professionals one was to please or receive praise from, but paid agents of a system that owed them. Thus, resisting the "system" continued to be the mark of non-college-bound young white males:

> Warren was very intelligent but he fought the system. They would fail him because they didn't like him... He was fighting the system, wouldn't do his homework, wouldn't take the tests. He was an athlete also. We both did [fight the system].

College-bound jocks Jim and Josh, on the other hand, while they too coasted through their high school years with little effort invested in their studies, enjoyed generally peaceful relationships with their teachers. Solidly middle-class white, they moved in circles where their parents socialized with their teachers and where the parent-teacher connection easily morphed from a neighborly relationship to a political alliance that favored the parents' child. They remembered:

> Most of the relationships [with teachers] were pretty good. I knew some of them... a lot of the teachers lived around here.[29]
> A lot of the teachers you knew through your family... your parents knew them. I think it made it easier on us... The teachers knew our parents, they had to pass us.... Like cutting a girl from the soccer team. If you knew her parents really well, you might keep her on the soccer team.[30]

While all jocks held high profiles and enjoyed special privileges, it was jocks like Jim and Josh, the privileged among the privileged, who fully belonged, and for whom the "system" was not a problem. Their recollections echoed those of white middle-class female graduates of the Divided Generation, whose relationships with teachers were intimately interwoven with their family and social lives outside of school. For white students of means across the two generations, high school was an extension of home.

All of the jocks interviewed remembered loving and respecting their coaches.[31] But white middle-class Josh White, a 1976 graduate, also remembered the coach's rougher side and the jock culture's acceptance of the coach's display of physical force as a mark of privileged attention:

> Everybody liked him [the coach], but you were kind of afraid of him too...was rough on kids too...getting beat up by the coach was like a status symbol. You were proud of it.[32]

Alumni stories further suggested that male teachers in general privileged male jocks by taking particular interest in their game wins and losses, by cheering for them, and by pitting one sport team against another. Tim Whittle explained:

> [Jocks were privileged] especially when you got the male teachers...One social studies teacher was hysterical...If he liked the sport that you were playing, you were in...On Monday, he would bring it up in class. If a kid he didn't like was on a team that lost, he made sure to bring that up [too].

Male teachers discussed the jocks' performances during classroom hours, a practice that underscored the cultural weight that sports carried among males across teacher-student boundaries at Miller High. The male bonding around sports within the school was further highlighted by contrasting recollections of female teachers' particular scorn against jocks who did not attend to their studies, as earlier discussed.

Interracial Female Jocks

Male bonding notwithstanding, female jocks, while they "fell into the jock, semipopular group,"[33] and held less status than their male equivalents, also enjoyed more privileges with teachers and a higher profile among peers than students who were not involved in sports. African American alumna Teresa Randle, a 1981 graduate, explained: "I felt that if you were on a team, they [the teachers] worked more with you, because you had to leave early a lot of times."[34] Teresa and white alumna Sophie Baker, a 1985 graduate, fully and emphatically remembered their participation in high school as jocks. In no uncertain terms, Sophie stated: "I'm definitely the jock."[35] Teresa remembered: "I did field hockey in the fall, basketball in the winter, softball in the spring."[36] In keeping with their semipopular group status, female jocks' games were always strategically scheduled as warm-up acts for male jocks' performances:

> They would place the girls' basketball games before the boys' games on Friday evening. If you saw a rise in attendance in fourth quarter, it's because they wanted to get a good seat for the boys' game. There wasn't much attendance; they just started letting the girls have cheerleaders at some of our games.[37]

Regardless of the lesser attention girls' sports drew, female athletes played, as did their male counterparts, for the sheer love of it. Similar to male jocks' recollections, and in particular to David Randle's testimonies as a black man, alumnae remembered the long-lasting bonds created on sports fields.[38] These bonds, forged on the field between African American alumna Teresa and her teammates, opened up doors to relationships with the white girls on her team that gave her some social advantages:

> I didn't drive when I was in high school. I could always call on one of the popular girls in the clique to come and pick me up, to go to a party, so I never had a problem with transportation.

These associations also exposed her to intergenerational rifts among white people at the time regarding race relations. Teresa recalled:

> I remember a friend, her father was a member of the Klan. I remember one time we were out and her car broke down. We had to go to her home. I was afraid...she said he wouldn't bother me. I was like, "Deb, I'm afraid to go into your house." "Oh, no," she said, "he's asleep. You'll be fine." I went in there with the four of us, and her mother brought us home. I was scared as anything.

Teresa's recollection of her white friend's split life, where she moved daily from the world of school and friendship with a black girl to life at home and a racist father, echoed David's and Tim's recollections of rifts among whites regarding race relations. Teresa further experienced these divisions among whites when she traveled with her white friends. She recalled:

> I used to go on vacations with my white friends. I'd go to an [ocean resort], never been to an [ocean resort] in my life. They introduced me to the [ocean resort]. You would hear little racial slurs as you were walking down the coastal highway. They were like, "I'm sorry." I didn't like it, but it was not something I was going to fight...At that time *Roots* came out. I remember going to school and everybody saying I'm sorry they treated you like that

back then...They were all very nice...Nobody treated me poorly at their home, eating dinner.

Recollections painted the world of sports as a place where students crossed racial divides and black students in particular sealed long-lasting relationships with their white peers. Black and white alumni testimonies across almost three decades of high school life since desegregation at Miller High in 1956 continued to accord sports a prominent place as where black and white students bonded within their male and female groups.

However, although in the 1970s and early 1980s many white peers felt responsible for and protective of black peers, by 1985 white students' attitudes toward their black friends were less apologetic and less constructed around the fragility of black/white relations. Sophie explained:

> The black people I disliked, it wasn't because they were black, [but] because they fell into this other group [potheads]. There was one of them that harassed girls in the hallways...He would pinch girls going by, stuff like that. We all pretty much hated him.

Throughout the 1970s and early 1980s, students' relationships across racial boundaries grew progressively more comfortable. By contrast to students of the Divided Generation, when everything "was definitely black and white," students in the 1970s and early 1980s seemed to base their feelings primarily on shared interests and values, and not on race or class. Thus Tim, a white jock, was friends with Teresa, a black female athlete, while Sophie, a white female jock who married and African American man, and her jock friends could not stand the likes of one black male pothead.

Graduates' recollections also revealed that some male jocks' notions of girls were challenging the cheerleader/jock stereotype. When one male jock remembered the athlete girls, he described them as earthy and, by contrasting them to cheerleaders, whom he ridiculed, gave them a higher status. Jock Jeremy Garnes, a 1978 graduate, remembered:

> The girl athletes back in our time, they were great. The cheerleaders, the prissy girls who didn't want to get their fingernails dirty and that kind of stuff! The girl athletes, they were just down to earth.

Jeremy's recollections of his female counterparts suggested perhaps less of a consensus among peers, and in particular among male jocks, regarding the desirability of cheerleaders. Conversely, the female jock was not completely without vanity. For Sophie, part of the allure of being a jock was the possessions that came along with the status. She fondly remembered:[39]

> I was just glad to make the varsity team because you get the jacket with the letter, with the *M* and the pins...I still have it, it's here in the house in the attic. I can't fit in it. I can't part with it. I just look at it.

Both African American alumna Teresa and white alumna Sophie "took it for granted that you needed a diploma and that it was a must." They never doubted that they would graduate. It was expected of them, and they expected it of themselves. As for the ceremonies, while Teresa reveled in them, Sophie could have done without them:

> Teresa: I was excited...It was a big accomplishment, I made it. We had a party; everybody was really excited.
> Sophie: I didn't want to go to graduation because it was this big ceremony, cap and gown...a long drawn-out calling of names, it seemed...I just wanted to go to the beach.

Although both were college bound, neither applied themselves as a student. They estimated the minimum effort required and delivered accordingly.[40] Regardless of their admitted lack of effort, neither alumna thought that Miller High provided them with a good education, echoing male counterparts' recollections. Teresa and Sophie recalled:

> Teresa: Actually, [Miller High] did not give me a good education. I think at that time, I didn't know any better.
> Sophie: "I didn't form study habits. I wasn't prepared to write on the college level."

However, they did not attribute their poor education to teachers directly, but to their own lack of effort and to the broader pervasive lack of standards that made getting by an easy thing to do, implying perhaps that a more demanding academic structure might have kept them on track.[41] Unlike their male counterparts, they assumed individual responsibility for their lackluster performances, and unlike David and Tim, both alumnae reported having good relationships with their teachers.

Of interest is that when Teresa explained her relationships with teachers, she used the term "indifference" as opposed to "different treatment," the term that had been used by African American alumni of the Divided Generation:

> In my perspective, I can't say I had any teacher treated me indifferent. That's actually from first grade on down, because I'm born and raised in this area. I went to Miller Elementary, Miller Junior High, Miller Senior High.

When African American jock David remembered teachers' relationships with black students, he too alluded to "indifferent" behavior. The earlier generation of black students at Miller High constructed cross-racial relationships more in terms of just or unjust, equal or unequal treatment of all students. Perhaps the only expression of inequality that could not elude the radar of discrimination by the 1970s was *indifference*. Unlike David, however, Teresa remembered her own relationships with teachers favorably and further attributed her good relationships with faculty to her deep roots in the community. She echoed the more affluent white jocks' recollections of life made easier by virtue of parents knowing teachers and teachers knowing generations of family members. Recollections suggested that black female athletes, already in a semiprivileged position in the school hierarchy by virtue of being athletes, might have experienced more favorable relationships with teachers than nonathletic black students in general, and nonathletic black male students in particular. Of note here is that David's recollections of teachers' indifference toward black students applied less to him than to his male black peers as he saw their athletic talents wasted. Teachers' indifference to African American non-jocks further reinforced the privileged position held by the athlete, male or female, black or white. Not unlike David, however, Teresa also recalled guidance counselors informing her only of all-black colleges and failing to mention other choices.[42]

Like David, Tim, Jim, and Josh, Teresa and Sophie remembered enjoying their high school years, which for them fully revolved around playing sports. Teresa happily recalled: "I truly enjoyed the...years I was at Miller High. I have made a lot of good friends. I still keep in touch with them. I just had a good time!"

Interracial, Cross-Class, Cross-Gender, Cross-Track Potheads

By contrast to the high-profile status held by jocks at Miller High, potheads were the "parking lot" students: "[T]hey were the people who smoked outside before they came to school."[43] Graduates' recollections revealed an animosity between potheads and jocks,[44] explained by lifestyle choices. Jocks saw themselves as healthier than potheads, suggesting that to them, drinking wasn't as bad as smoking pot. Jocks mostly "did not fool around with drugs," that was the world of the "heads."[45] Jocks drank. Jim and Josh recalled:

> Jim: "We did drink a lot."[46]
>
> Josh: "I mean, the jocks used to drink a lot...the drinking age back then was eighteen. We used to drink a lot."[47]

Jocks saw themselves as the achievers and saw the potheads as wasting their time. Josh said, "Our mailman, who was in high school the same time I was, he was a head. He admits to it. He was a mess back then. It's funny to see him now."[48] Josh's allusion to the mailman's admittance of having been a pothead at once captures present-day scorn against drug use, as well as the jock's construction of the pothead as socially inferior. Although there were also female jocks and female potheads, the rivalry polarized males of both subgroups. African American alumnus David Randle remembered:

> They [the potheads] had their group and we were the jocks... [There were rivalries] just with the jocks and the potheads.

While jocks overtly rejected potheads, as did many teachers, who called them "losers,"[49] potheads accepted anyone and everyone who shared their affinity for smoking marijuana and getting high. Within their group, they included females and males, white and black students, musicians, achieving students, and even jocks, despite the rivalry. Students were regulars or visited on occasion. For Sam Garnes, a 1974 graduate and then a regular pothead, school was an unpleasant experience. Sam, who began smoking marijuana in high school, remembered his first day as a stressful one: "I remember being confused, being worried about making it to the right classroom at the right time."[50] Unlike his jock peers who received their schedules ahead of time and became acquainted with school premises during field practices a month before school started, Sam entered the world of high school a stranger. He "didn't want to be there." His days revolved around getting stoned:

> *Did people go to class stoned?*
>   All the time.
>   *Did teachers react to that?*
>   Nah. Not at all. Basically you'd get stoned in the morning, then at lunchtime, then after school got stoned again.

The rivalry between jocks and potheads recounted by the jock graduates was also recalled by Sam:

> There was the jocks, which were the sports people, and then there was the heads, people who smoked pot... It was the competition between the jocks and the heads... the jocks assumed they were better than everybody... and wanted to be recognized... Jocks stuck together basically. The heads hung out in the parking lot... [where] you weren't bothered by teachers.

From Sam's point of view, the jocks' arrogance divided the two groups. Beyond attitudes, the spaces claimed by each group further divided them into the more and less visible students. Jocks' visibility was pervasive. They were seen in the cafeteria, on the sports fields, and throughout school premises. Virtually all the time, they were *seen*. Not only would potheads not be seen during sport events, but they would also disappear several times a day into the outdoor parking lot, out of view. The cultural connotations surrounding the parking lot as a place where illicit behavior occurred[51] further separated the potheads from mainstream high school activities. In Sam's memory, the school was squarely divided into two groups: those who would "make it" and be successful and, on the other side, those who were considered the "losers,"[52] a concept communicated and reinforced by teachers themselves.

Throughout his high school career, Sam worked an average of twenty hours a week parking cars at restaurants, mowing lawns, and working at his father's laundry business. As was the case with many young white men of the Divided Generation, Sam visited school more than he attended it. Unlike the poorer white boys of the 1950s and 1960s, however, Sam was a middle-class suburban young man alienated from school life. Paradoxically, alienated as he might have felt, Sam believed in high school. Apart from being a place where he smoked pot with friends in the parking lot, for Sam, Miller High was also a place that anchored his life. Its usefulness resided in its predictability. If it taught him anything, it was a sense of contractual responsibility whereby he presented his body to the institution to be counted. Explaining the meaning of high school attendance, Sam said that is was "[j]ust the responsibility. Getting up and going to school, a routine for life." For Sam, academic tasks were meaningless—their rationales were not clear, and they offered no satisfaction:

> Homework, I imagine there was a reason behind that, to make you better at what you were doing, but I just didn't enjoy it.

Sam's description of homework as not being enjoyable introduced the criterion of pleasure in the construction of schoolwork. Similarly, Sam constructed teachers' efficacy in terms of their visible enjoyment or lack of enjoyment of their work as teachers, echoing other graduates' recollections of the Border-Crossing Generation regarding the importance of "liking" what one does. Sam explained:

> Some of [the teachers] seemed like they enjoyed their job, and some of them were just there to put the day in. You could sense it... they were boring, very strict discipline, no talking.

While students of the Divided Generation constructed schoolwork predominantly in terms of usefulness, and teachers' roles predominantly in terms of their acceptance and understanding of students' backgrounds, alumni of the Border-Crossing Generation added yet another dimension to their understanding of both work and teachers' roles: finding pleasure in the work itself, in the process, not just the outcome. A good teacher in Sam's opinion was "somebody that could speak to you, could relate to you...that had some different fun things [for you to do]."

Sam, a white suburban student of the 1970s, like Robert Heart, a poor rural student of the 1950s, "looked up to and respected" his shop teacher, in whose class he "was totally interested all the way." Fond memories of the shop teacher shared across generations of white male high school students who visited school more than attended it, who worked at jobs far more than they ever studied, and who, when it came to school, "didn't want to be there," suggested the importance of hands-on activities and concrete work.

Just as jocks reported that "getting beat up by the coach was like a status symbol" of which you were proud, Sam too shared memories of coaches' aggression toward young men and the respect their violence inspired.

> Shower was mandatory after gym. Back then, you used to take showers after phys. ed. Coach would smack you with the belt, whistle strap. You had respect for him. Tough, but you had respect. You would have welts, but you took it.

The image of naked bodies blistering with welts after a whipping administered in the shower stalls, a space where one is rendered fully defenseless, is likely to be suggestive within present-day perspective of abuse of power rather than respect for power. White male students' call for understanding on the part of their teachers on the one hand, and their acceptance of male school authorities' physical aggression on the other, revealed a conflicted disposition of mind and heart whereby tenderness was yearned for, but aggression was respected. This was reflected in alumni testimonies across time, and particularly those of white alumni who were marginal students. This duality—inherited, perpetuated, and transmitted across generations of non-college-bound white male students—brings into relief the emotional disconnect and physical punishment that they integrated as part of high school attendance. Making it through high school was a question of putting up with it, of doing time. While the shop teacher made it bearable for the disengaged, for non-college-bound young white men like Robert Heart in the 1950s, and Sam Garnes in the 1970s, graduation was not about skills acquired or plans for the future. It was about the end of an ordeal. On graduation day, Sam and his fellow potheads were passing a joint around in full view.

Not all pot smokers however, were considered potheads. Joanne Pet, an African American alumna and 1974 graduate, also shared Sam

Garnes' affinity for smoking marijuana. Joanne explained: "People did more drugs than they do now...the joint was pretty much the cool thing, more than drinking...'cause the alcohol left a smell...I was pretty crazy and pretty high myself."[53] Like Sam, Joanne did not participate in extracurricular activities, and was not "a school-spirited person." Unlike Sam, however, she "loved the atmosphere of going to school." Joanne made friends easily and established longtime relationships with white female students:

> I had no transitional problem as far as making friends and getting along...My friends basically were all white, and right now, my very best friend is a white lady. Very best friend. I could call her now and she would do anything for me, and vice versa. She's not my friend because she's white, or the fact that she's not black, but her personality, *our* [speaker's emphasis] personalities.

Although she had been "pretty crazy and pretty high," Joanne "did well in school...and had no problems at all going to Miller high." Smoking pot, for Joanne, was part of the high school social scene, not an antidote to classroom boredom.

Her goal was "to be a secretary and work in an office." Although Joanne's relationships with teachers had been good and devoid of discrimination, the reality that awaited her in the job market was completely different. In this sense, Joanne's recollections echoed those of white and black women of the Divided Generation who described the contrast between great hopes and promises of high school, and the discriminatory realities of the job market:

> So I took the job...This is when discrimination slapped me right up against the head. Knocked me down like a brick wall. And then I get there, and no black people work in the office...The black people work in the factory.

In this sense, for women in particular, the middle-class white women of the Divided Generation and the black middle-class women of the Border-Crossing Generation, high school continued "to provide...a social experience that [was] markedly more egalitarian and more open to free choice and possibilities of self-realization than anything that [was] available to them in the realm of work."[54] Graduating from high school, however, was a time of great pride for Joanne:

> That was a big thing. Had a graduation party...My brother came home from Ohio...and he drove home for my graduation, him and his family,

and that was the highlight for me, 'cause he didn't come home often... He
and I were the only ones who graduated from high school.

Sharing her achievement with the only sibling who graduated from high
school was deeply meaningful to Joanne, who grew up in a family that
paid little attention to acquiring an education. She remembered the messages she received in early childhood:

When I was a kid growing up in my house, it was kind of cut and dry.
"Oh, you didn't go to school today?" "Nah, tomorrow." "Hum. Didn't go
to school today?" "No, later." "Well, maybe you need to look for a job." ... It
was just like, "I don't want to go to school anymore." "Okay. Get a job."

Joanne's recollections challenged the stereotype of the pot smoker as
"loser" on several levels. Not only did she succeed in acquiring a high
school diploma against the odds in view of her family background, but she
also did not lose track of her desire to become a secretary, even though she
regularly hung out with pot smokers. Her ability to make friends easily
notwithstanding, Joanne explained her high school success, and indeed the
rewarding life that she has led, in terms of learning from others' mistakes
and poor choices. Joanne defined the term mentor in an unusual way:

A lot of people today look at mentors to be great people—Michael Jordan,
Oprah Winfrey, whoever. My theory is it doesn't have to be that way. Your
mentor can be the other direction. My sister, she got pregnant she was eighteen, got married, and the next day had another baby, had another baby,
bad marriage, low income. That was my mentor. Something *I do not want
to do* [speaker's emphasis].

While graduation was a time of great pride, it was not a time of hope
for Joanne, as it had not been for many black youth of the Divided
Generation. She recalled: "I didn't see any hope. I guess the only thing
I knew was you graduated from high school and you went to work."
Joanne's remembered lack of hope during graduation but pride in the act
of graduating revealed a mixture of stoicism, akin to that of working-class white young men proud to have made it through the system, and of
disappointment, akin to that of white women of the Divided Generation
whose experiences of success within high school would not be matched
in the job market. Thus, young women and men, white and black, for
whom college was not an option, as alumni recollections revealed across
the decades and into the 1970s, experienced high school as a personal
challenge. The diploma was the symbol of beating the system, or vindicating parents' background of poverty, or dissociating oneself from a

life of bad marriages and ongoing pregnancies. The diploma, for these students, symbolized not so much what they had achieved but what they had avoided, and perhaps in that sense did not offer much hope. The diploma attested to the fact that they were not failures rather than that they could be successes.

Interracial Male Musicians

Beyond jocks, potheads, and friends of potheads, some students identified themselves as musicians. This was the case of African American alumnus Pat Baley and white alumnus Michael Hallner, graduates of 1975 and 1976.[55] While in popular culture, musicians in the 1970s were stereotyped as drug users on a perpetual high, and while it is almost certain that many Miller High musicians were pot smokers, Pat and Michael were not "airheads." However, they shared with potheads such as Sam a peripheral involvement in school. Pat explained: "We [musicians] didn't try hard, we just got by. Michael echoed: "If I got Cs, that was good. The goal was to get through it, get done with high school."

For middle-class African American alumnus Pat and middle-class white alumnus Michael, playing music was more important than anything school had to offer. Getting by with the anticipation of getting out was how these garage musicians made it through high school. Their lives began after school. Michael remembered:

> When school was out, it was my time and that's how I felt about it. So it's time to go home and do my thing, you know.

The world of musicians that Pat and Michael recalled was solely a male world. In Pat's band, males were black and white. In Michael's, they were all white. Unlike male and female jocks and unlike Joanne Pet, adolescents who looked forward to going to school, Pat and Michael looked forward to leaving school and playing music in their garages. Their recollections echoed those of many white and black young men of the Divided Generation. To them, schoolwork seemed irrelevant to life at large—"It was dull."

For Pat and Michael, the "goal was to get through. It wasn't to excel, it was: you're here, just get through it, get done with schoolwork, get on." Like many young males, black and white, who stuck it through, neither considered dropping out. Michael never thought of dropping out for fear of having to contend with his parents' displeasure. Typing was the only class he remembered as being of some use to him in later life: "I learned more in that class that benefits me now than anything. Learning how to type was the best thing."

While Pat also just "got by," he saw the usefulness of a high school education in terms of social capital and the relationships that turned out to be important connections for economic success later in life. He explained:

> You got social skills [at high school]. You met people there that...down the road might own a business and you got a job 'cause you knew them.

Pat and Michael were both peripheral students who put little effort into their studies, did not participate in any extracurricular activities, and lived to play music after school hours. While for Michael, a middle-class suburban white young man, the whole experience was fairly useless, for Pat, an African American middle-class young man, high school provided the opportunity to seal friendships across racial barriers, friendships that proved to be useful connections later in life. For white young males of some means who graduated in the 1970s and early 1980s, as had been the case since the early decades of the twentieth century, high school had little bearing on their economic fortunes. Furthermore, the particular demographics and geographical configurations of Miller Town, which more readily juxtaposed richer and poorer, white and black, gave many black families in the neighborhood the opportunity to live side by side with wealthier white people. Pat recalled: "I grew up in a neighborhood where there was nothing but [white] rich people. Now I wasn't. But they were my best friends."

Other Interracial Peer Groups and New Attitudes across Race toward Pregnant Youth

Recollections brought to view other student-generated labels that identified the very smart and academically successful male students as "eggheads"; the mechanically inclined shop students, all males, whose passions were to work on cars as "motorheads;" and the "fireheads," all males again, who volunteered for the town's fire department and spent as much time as they could around the firehouse. Except for the labels "jock," "prep," and "pothead," labels were generally assigned to males: "The guys were eggheads, girls were just smart."[56]

Although young women continued to get pregnant at Miller High, they were less likely to leave school than pregnant students in the 1950s and 1960s. African American alumna Teresa explained:

> There was a pregnant girl, a white girl, she had the baby, gave it up for adoption, and came back to school. Another girl had a baby, kept her baby, and

came back to school, so she was a mother when she was in junior high; she was black. Then a couple of girls that walked across the stage on graduation that were pregnant. They ended up having the babies right after graduation from high school.

The Poor and Other Marginalized Students

Only two alumni remembered "poor" students whom they described as marginalized. Musician Michael Hallner recalled:

> The few that were poor, they turned out to be the ones that were picked on, not because they were poor, but the way they dressed. There was one girl, one boy...they were the same ones all the way through twelfth grade...almost every kid in the class would pick on them...In twelfth grade the girl was there, but I don't think the guy was still there.

Tim Whittle also remembered, but more harshly:

> The real poor kids were teased just because they were dirty...they didn't get teased because they were poor, they got teased because they were pigs. They kind of brought it on themselves.

Tim's particularly virulent indictment of the poor might have been a reflection of his own need to distance himself from poverty that lurked around the corner as he was growing up. Michael's and Tim's recollections depicted merciless behavior toward the poor identified also in recollections of graduates of the Divided Generation, suggesting a continuation into the 1970s and 1980s of the disdain for the poor, whose "dirty" appearance betrayed their economic status. Except for Michael's and Tim's testimonies, other graduates' recollections during this period did not bring into view the poor, even when directly asked about them. In fact, African American alumna Teresa, who graduated the same year as her white peer Tim, remembered only middle-class students: "Everybody was middle class. One or two working class, everybody was in the same class."

This generation's more relaxed appearance might have loosened the hold the clean look had during the Divided Generation, and might have made the "sloppy" look more acceptable, blurring lines between haves and have-nots. Alumna Joanne Pet described potheads in this way: "[Potheads] had really, really long hair, raggedy clothes, eyes red." However, the teasing of the very poor as reported by musician Michael and jock Tim might have been perpetuated by enough students for the two of them to have noticed and remembered. The very poor at Miller High would continue to be stigmatized as personally deficient; and, as in the 1950s and 1960s, the very poor would continue to drop out, as Michael's testimony quoted earlier suggested.

Another group of marginalized students came to view indirectly, as remembered by others. While these students did not count among the very poor, they were not going to college, did not participate in sports, and were part of the working class, as Tim Whittle recalled. They were treated differently, not by students, but by teachers.

> "I saw being treated differently were...your group that did work-release, that took half-days and went to work. It sounds like they were in prison. I didn't like the word "work-release." Most of these guys didn't play a sport.... Guys who weren't connected to school...and didn't want to be involved in school. Some of the teachers gave up on them, on certain kids."

These students lived their lives within a school sanctioned limbo-like structure, where they were neither fully students nor fully employees. The work-release students were strangers to the school and its students since they attended only half-days. They were denied the social interactions that, as alumni stories have consistently revealed, offered opportunities for amassing social capital, for practicing alternative identities, for imagining a future different from one's past.

The Preps

Of all the peer groups identified by alumni of the Border-Crossing Generation, one group of students, the "preps," continued to hold a highly visible profile. They produced the yearbooks, wrote for the newspapers, organized school dances, and continued to create the image of the school. Described through the eyes of African American alumna Teresa Randle and white alumna Sophie Baker, the preps of the Border-Crossing Generation resembled those that Linda Moss, alumna of the Divided Generation, described as "the doers...the people that were on the teams...were the people that were putting out the yearbook, were the people doing that, you know, class officers...."[57] African American jock Teresa Randle remembered the preps as "the ones that were the class officers, cheerleaders."

However, Sophie Baker's description of the preps is the first among alumni recollections to represent the "doers" in an unfavorable light. Sophie remembered the undemocratic actions of the preps, who wore "alligator shirts and docksiders," and how they imposed their choice of prom song against majority will:

> They were...in the SGA. They sent out ballots for what you wanted to be your senior prom song. It seemed everything we voted for never went. We would count how many people wrote [in the ballot] this song [that they chose]...how did this [other] song make it?

By the 1990s, as we shall see, many more voices would criticize the preps and boycott their orchestrated school activities.

A Border-Crossing Generation Pushing against Structural Forces

Rivalries among peer groups notwithstanding, the Border-Crossing Generation of the 1970s and early 1980s ventured more freely across class, gender, and race divides, and tended to submerge these categories under youth-generated identities, from potheads to musicians. This was overall an accepting generation. African American jock Teresa Randle remembered:

> Everybody was allowed to be their own individual self. I don't remember anybody being aggressive toward anybody. They just knew that they hung over here, and the other ones hung over there. It was a given.

However, while students of the Border-Crossing Generation interacted more democratically than had the previous generation, many still contended with structural forces that constricted their movements across racial and class categories, even as peer groups offered opportunities for trumping race, class, and gender, as I explore in chapter 4. In the chapter that follows, I also examine the changing meanings of race for white youth, and the negotiations of race by black youth.

Chapter 4

Bridging Social Divides Through Peer-Groups: A Socially Tolerant but Politically Inactive Student Body (1970–1985)

Miller High students, as earlier suggested, associated mostly on the basis of common interests and affinities within interracial, cross-track, and sometimes even cross-gender peer-generated groups. Describing the cafeteria, Tim Whittle recalled that "male jocks [black and white] had a table, girl jocks [black and white] had a table, the eggheads [black and white] had a table...," and so on. Overwhelmingly, graduates remembered blacks and whites getting along within and across the peer groups.[1] African American Joanne Pet's recollection captured the civil rights mood of the time:

> Given that it was five or six years after Martin Luther King and the civil rights [movement], I was right there on the crest of it. So I didn't feel the effect of racism...I don't remember any incident that I felt that I was discriminated against because I was black...My friends, basically, they were all white.

However, structural forces—namely a split, inconsistent, and at times grossly insensitive faculty—divided students into winners and losers, neglected whole segments of students, and, as we shall see, crossed boundaries of propriety with their charges. Additionally, a continually pressing hierarchy of academic tracks continued to constrict the range of possible relationships of Miller High students even as they expanded their relationships across racial, gender, and track divides. Structural forces notwithstanding, within the setting of an increasingly suburban town and a national atmosphere of expanding civil rights, the Border-Crossing Generation interacted more democratically than the previous generation, although it never directly protested the status quo.

Peer Relationships

New Constructions of Race Relations and the Role of Peer Groups

From the majority of white students' perspectives, as suggested by recollection of both white and black graduates, outsiders were not the black students, but the minority white racist students whose discriminatory behaviors would quickly be brought to a halt by white peers. Unlike the Divided Generation, when black students who were assaulted by white peers sought help from school authorities, black students of the Border-Crossing Generation found ready help from most of their white peers. There seemed to be a more deeply integrated sense of racial equality among white students of the Border-Crossing Generation. This deepened sense of racial equality on part of the majority of white youth in the 1970s was further evidenced in the schisms in racial attitudes within white families, as the new generation of whites sought friendships with blacks against parental wishes.[2]

Graduates' testimonies further suggested that peer-group affiliation counteracted broader societal inequities and discrimination since belonging to a peer group typically meant belonging to an interracial group. However, students who did not belong to a peer group continued to segregate by race and class. Membership in a peer group offered an identity to layer over raw race and class distinctions. Without the cloak of peer-group appurtenance, one remained exposed as a census data category.[3] As earlier mentioned, using the phrase "social class" to describe peer groups, Tim Whittle recalled:

> If you weren't in a social class and you were black, you sat with the black kids. Teresa sat with the girl athletes; Amy sat with the girl athletes. Rona sat with the black girls because she didn't play a sport.

In the cafeteria, black students who were not part of the jock group or any other peer group sat together as black students. Of course, the story is more complex. African American alumna Joanne Pet explained that it was "a comfortable type of segregation":

> You segregate yourself regardless. We as people do it today. It's just a natural thing...black people sitting over here, guys over here, girls over here...It's just a natural thing.

Joanne's recollection about black students sitting together at lunch out of comfort rather than because of racist attitudes on the part of white students was also echoed by African American alumnus David Randle, who, while he felt at ease as a jock "eating anywhere," remembered the cafeteria segregation as a question of feeling "comfortable":

> We had open lunch back then, so you could bring in music. So naturally we [the blacks] wanted to hear something, and they [the whites] wanted to hear something else. They didn't force us in any groups. Just felt more comfortable.

White alumnus Time Whittle also remembered what happened when one didn't belong to a peer group: "You went with who you were most comfortable with...you went with your comfort area, kids you knew when you were younger."

Thus, a segment of the African American student population was primarily identified as black by their peers, rather than as jocks, musicians, potheads, fireheads, and so on. Many black students kept to themselves, whether proactively by refusing to participate in the white culture, as David's recollections suggested, or by default, either because they were not recruited by coaches into the mainstream sports culture, or because, as in the case of many young black women who continued the previous generations' roles of caretakers at home, they could not participate in after-school social activities with their peers. This segregation of black students suggests perhaps that the idea that everyone intermingled was the construction of the privileged students, white and black, whose high profiles at school, and/or identifiable appurtenance to a peer group, made it easier to navigate social borders of race, class, and, in some cases, gender. Lack of opportunity to participate or join peer groups and the natural tendency to seek comfort with one's own resulted in segregation by race for a segment of the black youth.

But "work-release" students also segregated. Work-release students were usually youth of lesser means who belonged neither fully to the marketplace nor to their school, and who were looked down upon by teachers, as we shall see in greater detail. The segregation of some black students and of work-release students restricted their opportunities to establish or broaden social connections. African American male and female jocks David and Teresa, African American male musician Pat Baley, and African American alumna Joanne Pet expanded their social networks within interracial peer groups, and forged interracial connections that continued into adulthood. Lacking peer-group affiliation, one became, by default, a member of communities that were racially and socioeconomically defined.

Overall, during the 1970s and early 1980s, male and female, white and black Miller High students along the upper- to lower-middle-class continuum experienced greater freedom to be themselves with each other, regardless of skin color and gender, than graduates of the Divided Generation ever had. Teresa Randle's testimony, however, of a visit to the house of a white friend whose father was a member of the KKK suggests a disproportionate burden placed on African American youth in relating across racial boundaries. Her testimony reveals the nonchalance of the white girl who seemed oblivious, perhaps by virtue of a combination of privilege and adolescent bravado in face of the status quo, to the terrifying implications for a black young woman of a possible encounter with a member of the KKK. It equally underscores the willingness and ease with which a white youth of the Border-Crossing Generation challenged the status quo, but again, with disproportionately little discomfort to her "self" in comparison to her black friend.

New Constructions of Gender Relations across Race and the Role of Peer Groups

By the 1970s, the dance floor was no longer a threatening place for black students. In contrast to the 1950s, when white middle-class teachers tacitly supported white students' exclusion of blacks at school dances, the 1970s and early 1980s witnessed an emancipation of allowable relationships on the dance floor at Miller High. Not only had it become a place where white and black couples danced side by side, but also a place where black males danced with white females and white males danced with black females. African American alumna Joanne Pet remembered:

> I had no problems looking at the white boys, thinking they were cute or vice versa. I had this guy...and the boy was drop-dead gorgeous, blond-haired white boy. And he asked me to parties. I mean it wasn't a question of am I going with him. What else am I going to do? If I'm living, I'm going. I'd have to be dead not to. You know what I mean?

Speaking about interracial couples, African American jock Teresa remembered: "The junior class, there was a couple. I can still see her, I can still see him...they are still married."

While students of the Border-Crossing Generation had substantially democratized dating patterns when compared to the Divided Generation, and interracial cross-gender relationships were now acceptable, new relational hierarchies emerged at Miller High, the result of the introduction of football at the very end of the 1960s. As several recollections revealed, jocks dated the cheerleaders and the best-looking girls in school. By the 1970s, cheerleaders were black and white. However, jock Jim Garnes' testimony about his preference for female jocks over the cheerleaders, addressed earlier in chapter 3, challenged the stereotype that male jocks are only interested in cheerleaders. Jim's memories of his female counterparts suggested less of a consensus among peers, and in particular among male jocks, regarding the desirability of the "cheerleader type," further suggesting that inherited binary constructions in popular culture around jock/cheerleader, popular/unpopular, or attractive/unattractive might never have represented the lived experiences of most high school students, but were the highly visible experiences of a very few. It might not be too bold to advance that "the distorted lens of historical mythologizing"[4] might be at play as movie directors, advertisers, and crafters of popular culture perpetuated and perpetuate certain stereotypes in high school peer associations.

Recollections also suggested that young women, when compared to young men, continued to be perceived as "definitely...a lot smarter."[5] From the 1950s to the early 1980s at Miller High, peers, and in particular male peers, whether white or black, continued to perceive females as smarter and better students. Female intelligence was also linked, across time periods, to effort, evidenced in graduates' references to young women doing homework for their male counterparts. The feminization of the "good" student throughout these periods, a characterization that shifted into the less feminine and more "purposeful" student in the 1990s, as we shall see, suggested an understanding of good students as those who did as they were told. Something that, at least as reported by graduates interviewed across the two generations, was better accomplished by young women than young men.

Youth and Authority

Constructing Teacher-Student Relationships across Social Categories

Unlike the Divided Generation, the Border-Crossing Generation depicted a student body less involved with faculty and school authority. For one, seldom did graduates, across race and gender, spontaneously remember their principals, and when asked about them, they vaguely recalled their presence, or not at all.[6]

Three principals succeeded each other between 1970 and 1985. The first two principals held office during 1970–1973 and 1973–1978. Although the third principal, Mr. Krauss, held office much longer, from 1978 to 1991, he was not remembered fondly by alumni who graduated before 1985. The very few graduates who remembered Mr. Krauss as an effective principal were graduates of the early 1990s, as we shall see, and for reasons connected to increased school violence at the end of the century. Recollections of the Border-Crossing Generation revealed an administration uninvolved in students' daily lives except when punishment was to be administered. Jim Garnes recalled:

> They [principal and vice principal] were distant. They didn't walk the hallways or anything... Only time you saw a vice principal is when you were in a lot of trouble. You were going to get yelled at. If you ever saw the principal, then you were in real, really big trouble. You really did something wrong.

While Mr. Lancaster had been beloved by Miller High students throughout his 1949 to 1964 tenure (and from 1956 on by both black and white students), as had been his successors who served during 1964–1966 and 1966–1970, principals in the 1970s and early 1980s were perceived by Miller High students not only as distant and punishing, but also as vengeful. Tim remembered: "This one vice-principal wanted [a buddy of mine, a football player] out of school."

Students' relationships with teachers, on the other hand, were not as clear-cut, and continued unevenly, as they had been for the Divided Generation. Students who butted heads the most with faculty were non-college-bound white males of lesser means, as Tim's recollections revealed. Some of these young men who either avoided or fought with teachers established connections with a coach if they were involved in sports (like jocks David and Tim), or with the shop teacher (like pothead Sam Garnes). The persona of the coach loomed large in the memories of black and white males for whom sports provided a place to belong. Some males in the 1970s seemed to accept and find status from being "beat up" by the coach, whether these young men were jocks or potheads. There is reason to believe that the introduction of football jock culture, nonexistent at Miller High before the very end of the 1960s, might have contributed to the greater acceptance on the part of many male students of coaches' aggressive behavior toward them as part of a manly show of stoic endurance.[7]

While coaches beat up on young males, other teachers literally made love to their female students. Narrators' recollections of teacher/student affairs in the 1970s and early 1980s brought into relief the transformation of a high school climate with the infusion of a very young faculty. Reports of student/teacher affairs that emerged in recollections of alumni of the Border-Crossing Generation were also remembered by graduates of the 1990s, suggesting a continued trend in teacher/student affairs for at least the latter three decades of the twentieth century at Miller High.

Sophie Baker recounted an affair between her friend and their teacher, an affair that escaped the radar screen of both school authorities and parents. Her recollection underscored not only the young friend's inner turmoil and vulnerability, but also the young women's fascination with carrying out a love affair with an older man, a fascination that helped Sophie be an accomplice to the illicit relationship. In hindsight, at the time of the interview, and from the current perspective of a mother of a teenage daughter, Sophie regretted her participation in her friend's affair. As a youth, however, she vicariously lived out a fantasy through her friend's involvement with the older male teacher, and intensified the experience by keeping it a secret at the request of the teacher.

> I'm embarrassed to say this. I helped her. He was younger, closer to our age, and when teachers...say they care about you, love you, and she was having problems at home, parents fought all the time, as an adult I can look back and see that she needed attention, and he gave her attention in the wrong way...I was a little bit enchanted with the whole older man thing...I would drive her to his van...he made us swear to secrecy...This girl had low self-esteem, and when you take advantage of that, that makes you a predator.

Tim Whittle also reported rumors of an affair between an African American teacher and a white student in the late 1970s that, although unconfirmed, sketched out the pervasiveness in young people's imagination of the time of the possibilities of teacher/student love affairs. Tim recalled:

> I got to Miller High a year late on a scandal with a teacher and a girl...I think '77. A black male teacher with a student white girl—that was a double thing. I don't know if she was pregnant or if he just slept with her. That was some kind of a scandal...he disappeared.

Evidence suggests that the influx at Miller High during the 1970s and early 1980s of younger teachers, who, closer in age to their charges, spent time with students outside of school hours and fraternized with them, created a social environment where teacher and student could readily become friends and lovers. This might have been further supported by the broader social and legal climate of the times where students' rights were on the rise and school authorities' paternalistic roles on the decline.[8] Sophie recalled:

> When I was in school...I spent time with people [school peers and teachers] on the weekends, it was platonic [with] the teacher...especially with sports, [spent time] at [teacher's] house, cookouts.

Young teachers, as alumni reported, were also more apt to put up with student misbehavior if it meant greater student participation in classroom discussions, while impatient older teachers routinely sent their charges to the principal's office.[9] These younger teachers not only loosened rules of engagement within the classroom, but also warned students ahead of time when administrators were displeased with their behavior. In essence, teachers snitched on more senior school authorities and acted more as peers to their students than authority figures. As one alumnus put it: "the younger teachers would tell you when someone was on the prowl for us," while the old teachers were more likely to "send you to the office."

Young teachers' disengagement from the establishment on the one hand, and their behavioral allowances on the other, contributed to blurring the traditional teacher-student boundaries for both teachers and students, in turn making student-teacher affairs easier.

However, the more empowering consequences one might have expected from greater closeness with teachers, such as greater participation on the part of black students who did not belong to peer groups and on the part of work-release students, and in general a more politicized student body, never materialized. The split between much younger and much older faculty, according to recollections, created a rift that focused teachers' attention more on other teachers than on students. Tim recalled that there was "a huge difference" between the young and old teachers. He said: "I don't think they liked each other. That is what cracked me up."

While in rare cases female students were reported to have affairs with male teachers, in general, female students' relationships with their teachers were far less disappointing or contentious than they were for male students who were not college bound. Alumnae in general reported good relationships with teachers.[10] By contrast, alumni fondly remembered mostly the coach and the shop teacher. For nonintellectual males, who constructed their expertise in terms of physical prowess, discipline, and teamwork in sports, or by designing, organizing, and producing projects in shop—hands-on activities that required physical and manual dexterity—the coach and the shop teacher were influential male role models. However, while the coach enjoyed relatively high status at Miller High, the shop teacher held the lowest position on the hierarchical totem pole of teachers and subject matter. Whatever satisfaction male students might have derived from building and producing concrete objects in shop, one might assume they also integrated the institutionalized low value that such activities held.

Male students who took shop or were part of the work-release program held a lower status in the minds of teachers. Teachers "liked the smart kids, and the jocks."[11] Miller High faculty communicated a bias against manual work and part-time school attendance through work-release programs. Students themselves, however, did not share in their teachers' bias, strongly suggesting that the privileging of certain types of knowledge-making sites (i.e., shop versus math class) was an adult construction no longer espoused by high school students in the 1970s and early 1980s, as it had been by students of the Divided Generation.[12] Teachers, on the other hand, arranged students in polarized fashion, either vertically, as losers or winners—"A trigonometry teacher... [she said,] 'You're going to be a loser'"[13]—or as insiders and outsiders, as in the case of work-release students.

Examining Miller High from student perspectives brings into question the institutionalized hierarchy of knowledge-making sites—courses that are offered but that are not equally regarded. For example, teachers looked down on shop when compared to the academic track. For many male students across class and race, from the 1950s to the early 1980s, learning was meaningful when they were engaged in manual work, or in making music. Yet many teachers devalued these learning spaces. Even when school authorities officially sanctioned an activity deeply rooted in a social-efficiency goal[14] such as the work-release program, teachers devalued participation in the program, de facto devaluing it as a legitimate site of learning for students.

The absent principals, the disinterested older teachers, the imprudent younger teachers who snitched on administrators, and the feuds among faculty along the generation gap combined to weave an inconsistent and often unresponsive environment for students who did not count among the smart and the jocks. Thus, students of older teachers did well if they behaved, whether they studied or not. Students of younger female teachers did well if they were smart, whether they behaved or not. Male jock students of sports-minded younger male teachers did well even if they weren't smart. One can only imagine the myriad contradictory messages that young people sorted through and around which they negotiated their daily student lives. One can also imagine how such adult behavior would create in the eyes of young people a laughable high school world at best—"that is what cracked me up"—and at worst an alienating world in which they perceived that teachers "gave up on you."

Ironically, teachers accorded young people varying degrees of legitimacy as high school students depending on their peer-group affiliations. The affective importance attributed to group belonging, by students and teachers, is a phenomenon perhaps unique to the U.S. high school—a phenomenon that paradoxically speaks more of individuality than of community or group belonging.[15]

Students of the Border-Crossing Generation who did not belong to peer groups acquired racialized and class identities. Teachers, as recollections suggested, would lose sight of students as individuals when those students did not belong to a particular peer group. Thus, black students who did not belong to a peer group were robbed of "individual" status. They became the black kids who, as David's testimonies suggested, went individually unnoticed by the white faculty, something he lamented in view of the many talents he knew many of them to possess. David, on the other hand, by virtue of being a jock, was accorded individual attention. Work-release students who lived in a limbo state, neither full-time students nor full-time employees, were also robbed of individuality and were seen as the working-class people coming and going at the periphery of school life, a group whom many teachers "gave up on." This phenomenon, by which specific peer-group affiliations granted individual status and lack of specific peer-group affiliations erased the individual within racialized and class groups, did not automatically equate individual recognition with positive recognition. Potheads were usually identified as losers by teachers, as Tim Whittle recalled: "A buddy of mine . . . he loved to get high, the teacher said, 'I'm going to make sure you don't graduate high school, you're a loser.'" But it did accord individual attention.

Unlike recollections of black graduates of the Divided Generation, whose overall impression of school authority was favorable while relationships with peers were reported as problematic, black graduates of the Border-Crossing Generation experienced the reverse: a rapprochement with students, but a greater alienation from high school authorities. Nat Right, the African American alumnus of the early 1960s who was assaulted by a peer, remembered his principal fondly as a fair man and spoke of students finally getting to the "business of being students." By contrast, David, an African American graduate of the 1970s, remembered poor and disappointing relationships with teachers and administration, but very good, long-lasting relationships with white friends. In the 1970s and 1980s, "resisting the system"—which had been the modus operandi of many poor white males during the early days of desegregation at Miller High—was now becoming the modus operandi for black males who were not jocks.

Young black men to whom faculty and administration had not reached out and who could not find a space among peer-generated groups retreated into voluntary racial/cultural segregation within the comprehensive high school. They resisted the white system, as David's recollections revealed. This segregation, as earlier discussed, was not aimed at white students, but at the white school authorities. On the other hand, as yearbook data show, more young black women participated in school at all levels, and as discussed earlier, had better relationships with teachers in general.

Students' Social Liberalization does not Translate into Political Action

There is reason to believe that youth-generated peer associations within a generally tolerant student body where "everybody was allowed to be their own individual self" created social conditions within which peers were less likely to compare each other based on the academic tracks they attended. Why, then, did there not emerge a student-generated change in status allocation of subject matters or in content of subject matters? Not for the sake of quickly tucking away credits toward an easy credential, as education scholars have documented,[16] but for the sake of surviving an otherwise meaningless environment, as in the case of Sam and others like him for whom shop, for example, was a site of meaningful learning. Miller High students did not take the opportunity of their particular space in time to challenge adult-communicated hierarchies of learning sites, and instead retreated from confrontation with school authorities into their bounded or hybrid communities. Whether these were black non-jock males who refused to join the system, black and white musicians who lived for their music, or potheads who found solace in shop class, the opportunities for protest were there for students. High school students in other parts of the United States during the late 1960s and early 1970s, and in urban settings in particular, had already begun and continued to articulate their dissent against institutionalized inequalities. One such white urban student deconstructed the high school system as a reproduction theorist might have and articulated a desire for unity among youth across racial, class, and track divides.

> High school is used to put people in various slots. It puts black people or poor people into slots—they will be working class. People like us here will go to college and flounder around in the arts. We are not supposed to have any consciousness of what goes on in working kids' minds...kids are coming

to realize one another's needs and want to break out of this classification system.[17]

Similarly, African American alumnus David's disappointment with the only black teacher, who taught history, echoed that of city-dwelling African American youth of the earlier twentieth century, as documented by historian E.F. Frazier.[18] In *Negro Youth at the Crossroads*, Frazier quotes a disgruntled black youth whose feelings echo those of David: "I've often wondered why we didn't study more about Booker T. Washington than George Washington! No matter how much I try, I can never be George Washington." David's disappointment underscored the historically known dissatisfaction of black high school students with the white bias in the history curriculum and simultaneously revealed the absence of overt complaint by Miller High black students in the 1970s against such biases.[19] Young black men of Miller High who did not participate in sports and who were not academically engaged asserted their dissatisfaction by not participating in school life rather than protesting its curriculum.

There is reason to believe that the school's relatively small African American student population in the 1970s, the close-knit community where long-established black and white families shared an identity as Miller Town residents and within which African American small-business owners counted on white clients, and the one black teacher's reluctance or inability to galvanize the energies of Miller High black male youth toward academic participation combined to silence dissatisfied black youth whose families emphasized respect for authority and getting along with everyone. African American alumna Joanne Pet summarized other African American graduates' recollections of their upbringing: "I was familiar with getting along with white people."[20]

The lack of outreach on part of the white faculty to recruit black youth into sports or extracurricular activities, the one black teacher's failure to deliver the African American historical reality that might have motivated and inspired Miller High's black student population, and the force of parental influence that emphasized "getting along" contributed to keeping many black youth covertly dissatisfied and seeking solace in group rejection of a system by which they felt rejected. This was, however, a group rejection manifested in nonparticipation in extracurricular activities and in skipping classes, but not in dropping out completely or overtly protesting change.

However, while Miller High students, white and black, did not make public their possible critiques of institutional norms, they seemed to ignore adult-imposed academic and social hierarchies. Miller High students' attitudes toward various sites of knowledge making (i.e., classrooms, shop, sports field, work-release, etc.) were liberal and democratic, while teachers' attitudes were elitist. Overall, students arranged themselves on a horizontal continuum that juxtaposed them by labels that captured their passions and interests, rather than on a vertical continuum that divided them by scholastic or extracurricular accomplishments, even as teachers favored jocks and the smart kids.

Still, in reaction to a generally removed and unresponsive white administration and faculty, some black students abstracted faculty and administration into an all-encompassing and nebulous "white system," just as they might have felt abstracted as individuals into the overarching "black students." This was similar to the way that white males of lower socioeconomic backgrounds of the Divided Generation abstracted high school life into "the system," as they too felt themselves unseen and unrecognized.

The result going forward would be a hardened identification with race for many black students, and a hardened identification with the lower class for many white work-release students. These hardened divisions, as we shall explore in Part III, provided the foundation for a complete reorganization of student associations. The affect that sprung out of those hardened identifications, coupled with an explosive demographic shift and zero-tolerance policies delivered by increasingly distant school authorities, created the conditions for a rigid resegregation among students.

Part III

The Redivided Generation
(1986–2000)

Chapter 5

Memories of Segregation by Class, Race, Nationality, and Religion: Destabilizing Years of Shifting Demographics (1986–2000)

In the late 1980s and throughout the 1990s, Miller High students attended an overcrowded high school where it was "tough to get from one end to another [between classes]...If your class was next door, literally you had to fight through the hallway."[1] It was a school that by one alumna's recollection was "old and ugly."[2] The solution to the overcrowding came only at the very end of the century with the construction of a state-of-the-art 600-student addition that was completed in 2000.[3] An influx of new arrivals to Miller Town contributed to the overcrowding: immigrant Russian Jews[4] and ex-city dwellers, many of whom were taking advantage of the recently erected Section 8 apartments[5] and most of whom were black. During this period, the population explosion accelerated the expansion of Miller Town, as high-end and low-end houses filled up any remaining farm fields, restaurants multiplied, and grocery chains, gas stations, antique shops, shopping centers, and malls spread.[6]

This accelerated development and avalanche of newcomers created a reaction from both whites and blacks with long-time roots in the community toward all those whom they called "newcomers" and accused of importing bad city habits. Still, while a young Russian immigrant might hear an irate Miller High peer shout, "Russian go home,"[7] irritations by old-timers were mostly directed toward ex-city blacks who doubled the population of black students at Miller High.[8] Graduates explained the changes in Miller Town population in the 1990s:

> By the 1990s, the black population in Miller Town had changed a lot, because I don't like to say anything bad about it, but look at all the apartments... We didn't know those people, though. They were all newcomers.[9]
>
> In the '90s, that's when blacks that weren't raised here started moving here... This area just exploded. In the '90s, the population probably tripled. Farms were dropping left and right. Black kids from other towns were moving here, and they were different from Miller Town black kids, totally different. You ask the black kids from Miller Town, they will tell you. I will quote you my buddy C.J. [black friend], he said to me, any time a county black starts hanging out with a black from the city, the black kid is either dead or in jail. We have friends that are dead from high school because they started getting hooked up with other black kids from the city, then they became heroin addicts. I'm not saying it's their fault; it's just a different culture. Every kid I grew up with that was black, soon as he started hanging out with non-county people, black or white, they were in jail or dead... you're mixing two different cultures... that's when they had some racial issues... black kids from the city.[10]

In response to what many perceived to be city kids' nefarious influences on local kids, the blacks and whites who grew up in Miller Town and were now having families began sending their children to private schools or homeschooling them.[11]

In the last decade of the twentieth century, Miller Town transformed from a place where one was likely to cross paths with a known face into a town of xenophobic old-timers and eager newcomers hoping to find a better life. The town's high school was absorbing an unprecedented amount of newly transplanted youth "with attitude," among which a segment of the recently immigrated Russian youth, who identified themselves as the Russian mafia, trafficked in drugs, hated blacks and rednecks, and sported anti-American sentiments. "Rednecks" in turn, who as previous analyses revealed, had consistently displayed racist attitudes toward blacks, were no longer being silenced by vigilant white peers as they had been during the 1970s and early 1980s. To their list of black foes, those identified as rednecks now added Russian Jews, as we shall further explore.

No longer was Miller High nestled within an out-of-the-way rural-suburban town. It was now the high school of an urban-suburban town easily connected by a metro line to downtown city life. Miller Town had become accessible to visitors who could not afford to own cars. Black parents in the city desperate to see their children attend the better county schools concocted fictitious residential addresses and began sending their youth to Miller High, to the growing frustration of administrators who began investigating the legitimacy of black students' school attendance. Expulsions, previously unheard of, became more commonplace, as did undercover narcotic cops who were now looking for ecstasy more than for marijuana. One alumnus summarized the impact of social changes in Miller Town during the late 1980s and throughout the 1990s:

> I think perhaps the Miller Town area [was] beginning to realize that it's not isolated anymore, that it has to deal with the outside world, and certainly, seeing the growing pains that go with that.[12]

Miller High students were now taking classes called Education for Responsible Parenting; business classes included introductory courses in marketing; psychology classes had been added; and the popularity of lacrosse had replaced that of football. In the hallways, students kissing passionately were a regular occurrence.[13] Girls showed up to school wearing more revealing clothing than ever before in the history of Miller Comprehensive High School, and students who wore T-shirts advertising alcohol or drugs were called to task.[14]

It was also during this period when the rare young person who openly expressed being gay did so at the risk of great ridicule,[15] and when Miller High students participated in clubs that were religiously and racially defined, such as the Christian Youth Against Drunk Driving, the Christian Young Life Club, the Fellowship of Christian Athletes (FCA), the Jewish Youth Group, and the Black Awareness Club. A bible study group[16] met every morning. At Miller High, at the end of the twentieth century, young people of Judeo-Christian religious traditions openly congregated on school premises to profess their beliefs.

This generation of Miller High graduates inherited the previous generation's youth-defined peer groups, added the punks, goths, and geeks or nerds, and no longer talked about potheads but of drug users and dealers. However, unlike graduates of the Border-Crossing Generation, who perceived themselves and others as belonging to clearly defined peer groups—or to none at all, in which case they identified themselves and others by racial and economic status—graduates of the late 1980s and 1990s remembered themselves and others first and foremost as Christian, Jewish, Russian, black, farm girl, and once again as redneck, superimposing religious, national, ethnic, and class dimensions to youth categories inherited from the previous generation. These categories of identity, however, played out differently for rule-abiding students and those who challenged the system.[17]

Almost all alumni interviewed recalled a very regulated high school setting, depicted administrators as punishing forces, and spontaneously situated themselves vis-à-vis the school principal to underscore their rule-abiding status.[18] Alumni recollections suggested that *not* knowing the principal on a personal level was a gauge of one's consistent good behavior. Knowing the principal, on the other hand, could only mean that one had transgressed. Harry Rice, Sue Cohen, and Cecilia Hood, graduates of 1987, 1995, and 1996, respectively, when asked about their principals, explained:

> Harry Rice: "I cannot remember seeing the principal walk the hallways...was very strict. You stayed away from the office."
>
> Sue Cohen: "I couldn't tell you the name of the principal and vice principal...I couldn't tell you what he looked liked, I wasn't at the office much, I had no reason to be with the principal."
>
> Cecilia Hood: "Never dealt with [the principal], never had to."

Furthermore, some graduates referred to school rules as "the law":

> Mr. L., he laid down the law...If you got caught [using drugs] there was no questions asked, you were automatically expelled and off the school property...on the first day they told you, they set down the guidelines...no drugs, no hitting on teachers...Mr. L., he didn't ask too many questions. If you were caught...you were out the door. Cherry Gate (1988–1992), interview by author, October 23, 2003

Some spontaneously recited a litany of prescribed behaviors:

> There was no drinking, no smoking, no drugs, no weapons, can't bring anything to school that looks like a drug, don't lay a hand on a teacher, you don't hit a teacher, you're not supposed to fight...the basics. Heather Korn (1989–1993), interview by author, February 25, 2003

Of the three generations of alumni interviewed, this one stood alone in its vivid recollections of its school's disciplinary apparatus. Moreover, memories of graduates revealed increasing expulsions across the 1990s and decreasing "school spirit." Memories regarding strong community involvement and pride in being a Miller High student shared by those who attended in the late 1980s transformed for those who attended throughout the 1990s into memories of dim school spirit.[19] Alumna Heather Korn, a 1993 graduate, remembered: "The new kids that went to school behind me, they weren't as involved as we were, they weren't as excited about being there as we were." Testimonies further suggested that administration's hard-line approach to discipline contributed to the lack of spirit. Sue Cohen remembered how the "administration came down really hard":

> There was a tradition, a senior doughnut run, where we run across the street to Dunkin' Donuts...my year, if we decided to do it, we'd get kicked out of

school. The administration came down really hard my senior year...spirit day [when you're supposed to dress alike], nobody did that...nobody came to football games.

Rule-abiding students included those who belonged to religious youth groups; immigrant students seeking social mobility through lawfully sanctioned means; students deeply involved in community service occupations; and usually high-profile students with means, the preps, who continued as they had since before desegregation to craft the school's official representation to the outside world, although, as we shall see, with much less enthusiasm, and much less school-wide participation.

Rule Abiders

Religious Middle-Class Jewish and Christian White Youth

Sue Cohen, a 1995 graduate, belonged to the Jewish Youth Group, and Bill Jackman, a 1999 graduate, belonged to the Christian Young Life Club. Both Sue and Bill, along with friends who shared their beliefs, could be seen recruiting members to their religious organizations on school premises. Sue and Bill recalled:

> Sue Cohen: "We were always trying to recruit people for the youth group. If there was somebody I thought was really nice and Jewish, I would say, why don't you come to this meeting?"[20]
>
> Bill Jackman: "We'd try to get into the school culture...we'd talk to the principal...if you see us around, don't be scared, we're not here to hurt."[21]

Part of their ethos as young people openly committed to their religious beliefs and unapologetic about expressing them within the public school system was an avoidance of conspicuous consumption. Bill belonged to the Fellowship of Christian Athletes (FCA) and was part of a growing number of jocks who refrained from excessive drinking, or from drinking altogether. Within the Young Life Club, he played games once a week, learned some sign language, and listened to or gave "a five- to ten-minute Christian message at the end" of the club meeting. Sue underscored the Jewish Youth Group's adherence to sober behavior when she distanced herself from the stereotype of the Jewish American princess (JAP), which suggests conspicuous consumption and self-indulgence:

> Most of my close friends I had through my youth group... it was the Jewish Youth Group, it was really tight... There were other groups, the JAPs. I hate to say that reference, but they were. I didn't really socialize with them.

These graduates' memories also revealed a keen sense that high school meant little in and of itself, that it was just a "stepping-stone"[22] to get to college, and that it required a purposeful and disciplined engagement.

Sue and Bill attended a combination of honors and some "highly competitive" gifted and talented classes, saw themselves as rule-abiding students, and were fully engaged in extracurricular activities dear to their hearts. Sue invested her energies in journalism, which was "a big deal [to her] at the time," and where she "was the editorial's co-editor."[23] Bill played in the Lacrosse team, which "went to the regional finals a couple of years in a row." Sue also identified herself as "one [of the] nerds," and Bill identified himself as a jock. These were students of solidly middle-class background who understood the importance of investing their energies competitively during their high school years to increase their opportunities for acceptance into the colleges of their choice. Students of the Redivided Generation were very much aware that increasing numbers of high school students, whether they wanted or not, headed for college because it was "harder to get a job without a college degree... they [wouldn't] hire you as soon."[24] Sue and Bill remembered that teachers in honors and advanced placement (AP) classes were demanding,[25] and that competition among peers could be fierce. She recalled:

> The gifted classes were very competitive... A friend of mine was running for valedictorian and someone else was, too. That person took college courses over the summer, had it put toward high school, so that boosted that person's quality point average. The other kid didn't take another advanced course... It was really sad.[26]

These were dedicated students. Sue "cared about doing well and put a lot of stress on [herself]," and Bill "hung out with pretty high achievers." Just as they put the onus of disciplined and responsible behavior upon themselves, whether as young devoutly religious people or as academic achievers, they projected onto all other students the same onus of personal responsibility for academic success. In reflecting on Miller High students and teachers, Bill said that "if kids are willing to learn, they [the teachers] will teach them." Furthermore, for Bill and Sue, high school graduation celebrations such as the prom seemed superfluous. Bill remembered dances as the purview of girls who were usually more involved than boys "in a prom-type thing." Sue remembered them as devoid of any sense of closure: "A lot of people didn't go to the prom, and it didn't matter to us. We just didn't care, it was just school, it was a stepping-stone to get to college."

Their recollections further revealed that being a Jewish or Christian student first and foremost, and secondarily a nerd or jock, considerably narrowed one's world of associations. In an overcrowded school where it had become impossible to know most of your classmates, Sue and Bill, within the spaces of "tight" youth groups and learning alongside "pretty high achievers," lived their high school careers in particular streams of student life that ran parallel to many other streams.

First, they were less likely to intermingle with black students, who throughout the 1990s continued underrepresented in honors and gifted and talented (GT) classes,[27] and they were less likely to intermingle with nonreligious young people. When Sue recalled youth from other backgrounds, they were from another faith: "Christian people came to my bat mitzvah, I went to their confirmations; we all seemed to get along." They were also less likely to encounter sexually active youth or those who did drugs, let alone know them personally. Sue literally did not know whole segments of the student population:

> I don't remember any of my classmates getting pregnant...I heard other people did drugs. I didn't do drugs, so I didn't think [the drug-prevention classes] were useful...I didn't know anybody who had sex.

Bill, on the other hand, made it a point as a religious youth to distance himself from jocks who drank by participating in the Fellowship of Christian Athletes. Similarly, Sue purposely distanced herself from the JAPs—whom she could not completely avoid, since they shared the same secular culture—by actively retreating "into [her] own group" of religious Jewish youth.

While the Jewish Youth Group remained tight and included only young people who actively participated in their Jewish faith, the Christian Young People Club, as Bill explained, included a few Jews and some punks: "We had more than a few punks...and at least three or four kids that were Jewish." Still, the overwhelming majority of participants in the Christian Young People Club included athletes who were also part of the FCA and a lot of preps. Russians, who for the most part were secular Jews, as we shall see, and blacks never joined. In recalling participants in the Young People Club, Bill said:

> A large amount were athletes...and the preppy kids...We wouldn't get cross-racial; I guess there was a separation with the black and white side of issues in this area in general.

While many black youth were fully involved in their faith as Christians,[28] they did not enter the worlds of FCA or Young People Club. In fact, by the end of the 1990s, as Bill remembered, seating arrangements among students in the cafeteria clearly captured streams of students divided along ethnic and racial lines, and revealed parallel worlds that did not intermingle.

> I guess there was a separation with the black and the white side. Usually the black kids would hang out with black kids, the white kids hung out with the white kids, there was some crossover, not overabundance though... it wasn't like everybody that was black hated everybody that was white and vice versa. There were incidents [of racial discord] but not extremes. There were people in school that were racist either way, but it wasn't overall... There were the Asian Indian kids, not black, not white, race-wise, didn't fit anywhere... The Russian kids were definitely a group.

Immigrant Russian Youth

Among the "Russian kids" Bill alluded to was Ivan Strasky, a 2002 graduate. He began his testimony with a description that suggested an essential shift in the historical depiction of immigrant Russians of the turn of twentieth century:

> In the midst of Americans we always spoke Russian. You had to; that was the cool thing to do. Among ourselves, we spoke English.[29]

These were unlike the Russian immigrants of a century ago who struggled to speak "American" English, to present an "American" persona in public, and who in the privacy of their homes forsook teaching the mother tongue to their progeny. Russian immigrants of the 1990s often spoke English before they came to Miller Town, and continued to teach their offspring the mother tongue, fully aware of the advantages of being bilingual in any country in the twenty-first century. Many young Russians at Miller High used their bilingualism as an identity banner to signal their allegiance to "Russianness." Ivan and Vera, Ivan's elder peer who graduated in 1999, were among the rule-abiding Russian youth. These were strategic students for whom high school "was a means to an end." Ivan remembered two very distinct groups of young Russians attending Miller High and how he belonged to the academic group:

> I think there are two distinct groups of Russians. You have the highly academic Russians, immigrant Russians whose parents are always telling their kids, 'School, school, school, we came here for your benefit, we

sacrificed so you can reap the fruits of what we left behind, the only way you can succeed is school.' There are some kids who chose to take their parents' advice and listen to them...[and] others who chose the path of being defensive.

Ivan was an organized, goal-oriented, and highly motivated student who, along with other Russian youth like him who strove to vindicate their parents' sacrifices, single-mindedly pursued college-track courses and built relationships with teachers and guidance counselors along the way.[30] Vera also engaged in studies purposefully and successfully, and although her teacher-student relationships were formally impeccable, she confided that she could not even remember her teachers' names, dramatically suggesting that high school was a quick and swift passage on the way to long and serious studies in higher education.[31]

While Ivan and Vera attended mostly GT classes and immersed themselves in academics, they did not identify themselves as nerds or geeks, as did the white middle-class U.S.-born Harry Rice and Sue Cohen.[32] Ivan and Vera both solidly identified themselves as Russian immigrants. Recollections further suggested that others consistently identified Russian immigrant youth as Russians, not as jocks, nerds, and so on.[33] Thus, whether of nerdish or jockish[34] inclination, a Russian student remained a Russian student. The institutionalized nature of such peer groups as nerds or jocks made those labels U.S.-specific. A freshly immigrated youth remained a foreigner, at least for the duration of his high school years, and as such would not acquire an inherited peer-group identity. Ivan remembered how Russian students huddled together in the cafeteria: "Two tables with Russian kids, each table holds about twenty-five people, a lot of Russian kids!" Vera further recalled how in general it was much more comfortable to remain within her Russian group, where she felt at home, than to risk rejection by stepping into another group. Tired of trying to "fit in" throughout all her middle school, Vera spent her whole high school career alongside her Russian friends and boyfriends:

> All the Russian students hung out together...Actually, until ninth grade, I hardly hung out with Russian kids, I tried to fit, I spoke only English, drove my parents crazy...we spent so much time in middle school trying to fit in, that by the time we got to high school, it was we didn't want to fit in...some of us came when we were twelve or thirteen years old, it was easier to speak Russian...so it was pretty much all Russian people, Russian boyfriends, Russian friends.[35]

Like the GT students Sue and Bill, Ivan and Vera interpreted other students' lack of academic success as a question of personal choice and effort.[36]

However, as Ivan saw it, a particular type of immigrant student was *more* likely to succeed if he applied himself:

> My perspective, it is so much easier to be a Latino immigrant. You don't have to be that hardworking or that intelligent to get here. In the general population that is Mexican immigrants, the really smart ones are the average ones. The Russian and Indian immigrants, people that had to cross the ocean, and had to deal with governments, the iron curtain, it was parents who were witty that found their way through that.

Ivan understood the immigrant story as one of natural selection that privileged, in this case, his national/ethnic group. The initial "wit" required to transplant lives overseas, translated, as he saw it, through "genes and observing their parents" into producing more capable Russian youth. He attributed the fact that one "wouldn't find many Latinos in the upper structure" to the dual role played by genetics and role-modeling. Ivan's use of "overseas hardships" as supporting evidence for his assertion, however, suggested that he was not aware of the oppressive government regimes that many Latinos were fleeing or the life-threatening situations they endured on routes more tortuous than a direct flight from Europe. In his mythologizing of his immigrant story, Ivan did not include that it is precisely the Soviet liberalization of emigration policies of the late 1980s and early 1990s that allowed for the Russian Jew exodus at the end of the twentieth century. As is perhaps often the case for children of immigrants, who may feel obligated to vindicate their parents' sacrifices and who were raised on hearty helpings of hardship stories that tell of all that was endured for their benefit, Ivan constructed his immigrant story as an epic—one he also constructed for Asian Indians, among whom he counted his good friend Prag, who would go on to MIT after graduation.

Ivan proudly recounted that at Miller High, "the achievers are immigrants," Russian immigrants in particular who counted among the "really smart and hardworking Russian kids." While he did remember a couple of Russian athletes, the rest were "that pole of the Russian community [at Miller High] that did nothing extracurricular." It is "that pole" of Russian students who challenged the system and its "American" nature. Whether academically dedicated or anti-American and rebellious, most Russian students dealt with yet another identity layer—that of being Jewish.

Ivan and Vera shared the complex relationship that many Russian students developed around the "Jewish" label when it referred to the religion and not the culture, and they found themselves defending Judaism without knowing much about it:

> Ivan: "If Judaism was ever insulted, Russian kids immediately perked up their ears. The irony in that is that out of the group of Russian kids, none of them that I remember were ever bar mitzvahed. I was culturally Jewish, not by ceremony. Russians in Russia, they were never allowed to practice their religion. They called themselves Jewish; they often didn't have a clue of what the holiday is about... The Judaism has been pulled out of them. They come here, they can practice Judaism, they don't know what or how to do it... The Russian kids are called Jewish because they are Russian, and because they are so loyal [to Russianness], they have to defend Judaism."[37]
>
> Vera: "I didn't fully understand that I was Jewish until I came here, so in my family, we never got into religion."[38]

The label "Russian Jew," doggedly adhered to and flaunted by a segment of Russian students who preferred the mafia world to that of academia, as we shall later see, was often used, as Ivan remembered, like a shield against perceived discrimination by teachers. Ivan explained:

> Whenever the teacher challenged them, not because of being Russian, but because the teacher was being a teacher—"Why didn't you do your homework? Why are you late?"—they would automatically be defensive, and often perceive it as discrimination.

Unlike their defensive peers, and as successful students, Ivan and Vera sported the Russian immigrant label differently, not as a target of ire and reason for defensiveness, but as a vindication of their parents' strife. Vera explained:

> I knew I was going to college for sure, my parents were going to murder me if I didn't go... I graduated with honors. My parents were proud.

Religious beliefs or cultural identification with religion and strategic academic competition grounded the daily behavior of Sue Cohen, a U.S.-born Jew, of Bill Jackman, a U.S.-born Christian, and of Ivan and Vera, immigrant Russian youth who identified as Jews. These high achievers were not the only rule-abiding youth. Many "middle of the road" white students, who were neither members of youth groups nor high achievers, and who did not identify with a particular nationality or ethnic group, but who attended the lower general tracks and more broadly shared a lower socioeconomic status, welcomed zero-tolerance rules that, as they saw it, kept things orderly and safe for the majority.

White "Regular" Students

In describing their principals, Cherry Gate and Heather Korn, graduates of 1992 and 1993, respectively, remembered:

> Cherry Gate: "Mr. L., our principal, he laid down the law and, I mean, you were never upset going through the hallways."[39]
>
> Heather Korn: "Our principal, she commanded respect... she held her ground and I really liked that."[40]

These were students who took standard courses and whose classroom realities sketched out differently than for those attending the exclusive worlds of honors and GT courses. Bill Jackman, who attended one or two general-level courses and could compare them to his GT and honors classes, explained how "in the standard level, it was a little rougher... The teacher was more afraid, maybe not of students doing something to her, but just from keeping the class from getting too disorganized."[41]

As had been the case for previous generations, it is also within the "standard-level" courses through which the majority of the student body traveled. However, students of the Redivided Generation who lived in the lower academic tracks were more likely to encounter each other less as competitive academic performers and more as representatives of diverse racial, ethnic, cultural, and even class interests and struggles. It is therefore as black or white, Russian Jew or "redneck," subsidized apartment or middle-class suburban home dweller that some of these students competed with each other, and not as college-résumé builders.

It is within the standard-level courses where much of Miller Town's social wounds flared up. It was a place where young people lived closest to the economic realities of their town as work-release students, and as part-time emergency medical technicians (EMTs), firefighters, burger flippers, veterinary technicians, waiters and waitresses, cashiers, tellers, and employees at the many retail stores and corporate giants, from Target to Kmart to Home Depot, that by the 1990s had spread around town. It is also within the standard-level courses that one was more likely to come across drug dealers and witness illegal behavior.

Recollections of graduates of the Redivided Generation revealed growing incidents of violence. However, while those students who attended GT and honors classes barely remembered fights and only heard rumors, as did their counterparts of the Divided Generation, alumni who attended general classes reported feeling the impact of violent incidents more vividly. Heather Korn recalled how "you knew who the drug users were" and that "if you wanted to get drugs, you could have them in ten minutes." She also recalled a day when she was at home sick: "One of the older students brought a weapon to school, a really big gun. We saw it on the news. I was so glad I wasn't at school that day."

To Roberta Jones, Cherry Gate, and Heather Korn, graduates of 1991, 1992, and 1993, respectively, zero-tolerance policies were a welcome protection against what they remembered as increasing school violence. As students attending standard classes, they were more likely than their GT and honors counterparts to witness fights firsthand, in particular racial fights, which increased throughout the 1990s. Cherry Gate remembered, although furtively and regretting having shared her memory almost immediately after uttering it, how her own sister had been the target of racially related death threats:

> A few years after I went to school, yeah, my sister was supposed to graduate back in '96, and she actually got a death threat and she went to Springfield High. She moved because she was scared to go and stay at the same school because at that point in time it was the racial fights going on.[42]

While Cherry, who graduated in the early 1990s, suggested animosity from black youth toward her sister, Bill Jackman, who graduated at the end of the 1990s, spoke matter-of-factly about white as well as black racism, suggesting that as the 1990s progressed and racial and ethnic tensions escalated, many Miller High white youth considered "racism" no longer singly the sin of the whites.[43]

Still, Bill, who attended during the second half of the 1990s, the years that Cherry suggested were marked by "racial fights going on," and who attended mostly honors and GT classes, recalled racial fights as being incidental and far from representing the norm. The disparity in Cherry's and Bill's assessments of the severity of racial problems at Miller High could be interpreted, as earlier discussed, in terms of Bill's attending the removed spaces of GT classes, which made him less likely to witness violence among peers than if he attended standard courses. However, there is also reason to believe that Cherry's sister's involvement in the racial incident she recounted might have colored her interpretation of the extent and nature of racial problems at Miller High. While her testimony implied that the onus of the racial problems in her sister's case lay squarely on the threats from black students, it wasn't until almost a year later when I interviewed Cecilia Hood, who graduated the year Cherry's sister would have had she not gone to Springfield instead, that I reframed Cherry's testimony in terms of possible racist animosity by Cherry and her kin against newly arrived blacks. Cecilia recalled:

> I think twice in the course of my high school there was a big racial fight between the black kids and the farmers... Two kids from Miller High were sent to Springfield High, got expelled, and two others got expelled and were not allowed back.[44]

Cecilia's identification of the whites involved in the racial fight as being "farmers" fit Cherry's self-description as "a back-home-farm-country girl" who "worked at [her] parents' farm."[45] Cecilia's memory of two kids being sent to Springfield High matched Cherry's mention of her sister going to Springfield High. Their accounts differed only in that Cecilia remembered that the students had been expelled and sent to Springfield High, while Cherry's recollection implied that her sister had chosen to go to Springfield High because the racial tensions at Miller High had become untenable. Finally, Cherry's refusal to elaborate on the incident or to provide access to her sister further cast suspicions on her sister's total innocence in the racial incident. Then, too, it's quite possible that Cherry did not want to provide contact information for her sister simply because her sister harbored such bad memories regarding her transfer (or expulsion). Confusion about Cherry's sister's involvement in racial disputes notwithstanding, Cherry's and Cecilia's conflicting testimonies illustrate the escalating reality of racial tensions at Miller High during the 1990s between "farmers," as they called themselves, or "rednecks," as others pejoratively called them, and blacks as well as Russians.

Beyond their shared feelings of security under principals who "laid down the law," Cherry, Heather, and Roberta placed themselves somewhere between "the group that did drugs" and the "higher group... that you knew were really smart."[46] These alumnae remembered themselves as unassuming, down-to-earth students who hung out with the "standard" group.[47] These girls distanced themselves from the preps and those with money and constructed the distancing as a matter of moral choice. Cherry recalled:

> I never hung... I guess with the popular group because I didn't believe in their status, the way of doing, the way of thinking. It was about the right clothes, the right makeup, the right hairdos... they were up on the *Vogue* and stuff, you could feel they were snobs.[48]

While memories of Divided Generation alumnae highlighted the importance of "looks," and while memories of the Border-Crossing Generation alumnae played down the importance attributed to "looks" in general, the memories of these 1990s alumnae stressed the emotional tensions that young women experienced around issues of clothing as they reflected on one's economic status. The identity of the "prep" had been synonymous with upper-class status and high visibility in all social affairs pertaining to school across generations. However, recollections of alumnae of previous generations did not in general linger on the preps. One alumna, Sophie Baker, who graduated in 1985, at the very end of the Border-Crossing Generation, did remember the preps negatively, as earlier discussed. By the late 1980s and 1990s, the prep was not only openly criticized, but also avoided.

In Cherry's and Heather's eyes and those of their friends, the preps were vain people, exclusive rather than inclusive. By contrast to preps' undemocratic attitudes, Cherry and Heather saw their ways as solidly down to earth. They knew themselves as "regular people" who considered everybody as equal. Cherry recalled:

> My friend was also the same. She was middle class...she stayed away from them [the preps], she felt the same thing I did, and the groups that were just basic regular people. You could feel that they [the preps] were snobs. In other words, they would make you feel that. Okay?

Heather also recalled: "They had the snobby attitude, but they were no different than you and I."

There was yet another group of students that Heather in particular found offensive—to the senses: the Russians. Echoing the late nineteenth and early twentieth century "American" complaint against many immigrants' "poor hygiene," Heather recalled: "We had a large group of people who were not clean people, like they didn't shower."[49]

For these young women, who attended mostly general courses and worked in the community after school,[50] who felt shunned by the preps and unsafe without rules to protect them against students apt to engage in illegal drug dealing and violent behavior, high school graduation *was* the end of their student career, at least for the immediate future following it.

Cherry, Heather, and Roberta constructed their high school days not as "a means to an end" or a "stepping-stone to get to college," as their peers in GT classes might have, but as a process of maturation with graduation itself marking the end of the life stage of adolescence. Their recollections suggested young people who had fully integrated a sense of themselves as incomplete humans on the way to maturity, and had associated the idea of high school graduation as a marker of their new identities. While Roberta Jones lamented having to grow up, Cherry

Gate celebrated the process:

> Roberta Jone: "It's really hard being a teenager, because you're in the middle...still being parented by your parents, and it's hard now [at graduation] you know you got to do something."
> Cherry Gate: "I grew out of my shell into a better person."

These alumnae's recollections echoed to a certain degree those of the Divided Generation who attended the general or commercial tracks and who saw the end of high school as a rite of passage and the high school diploma as a legal tender. At graduation, one had the right to assume one's own independent life without stigma. Heather and Roberta did go on to community college; however, Heather, right after graduation, and Roberta shortly after, found that their high school diploma limited their job options and, along with Cherry, remembered experiencing the end of high school as a place where one left as an adult ready to assume adult roles.[51] The "standard classes," however, as Roberta's recollections suggested, were also the place to which young people went who didn't like school and who didn't know what their lives were yet about.

> I let everything interfere with my schoolwork. I didn't like it...I wasted all this time on something I didn't really enjoy...I just don't do well in school[52]

For them in particular, school was a no-win situation. Neither staying in it nor leaving it brought any sense of relief. Leaving it meant that "now you know you got to do something," yet you might still not have known what you were supposed to do or be outside school walls. Roberta recalled: "It [high school] still didn't help me pick out what I wanted to do." Neither ready to assume a place in the job market nor headed for college, and feeling as if she had wasted her time within high school, Roberta could not find a socially defined acceptable space within which to place herself. At graduation she was an adolescent required to be an adult in a world for which she felt unprepared.

Having spent their high school years closer to the job market than their GT peers, closer to girls who got pregnant and boys who enlisted, closer to drug dealers and social prejudice, it is little wonder that these students would have understood the end of high school more in terms of entrance into adult life than their GT peers, whose dependence on parents would have to be prolonged as they pursued undergraduate and graduate studies. For the standard-class rule-abiding students, high school still held some sense of finality. Many of these alumnae's male friends were part-time firefighters, emergency medical technicians, and sometimes work-release students,[53] young people whose lives daily and weekly were involved with the community. Inevitably they would soak in much more directly the effects of economic competition and social pressures, and integrate, perhaps even perpetuate, the social prejudices of their surroundings.[54]

The Preps

By contrast, preps' experiences of Miller High in the late 1980s and throughout the 1990s were dramatically different. Preps continued to be identified by their economic status and the high profile they assumed through their pervasive involvement in the school's extracurricular and social life. In describing the preps, Bill Jackman accentuated their wealth and isolated upbringing:

> Kids that knew each other their whole life, they grew up together, a lot of those kids played sports, that connection... they do have a decent amount of money... they did the same things growing up and people don't like to change... a lot of them dressed preppy, I think it was more they were used to hanging out with each other.

Russian immigrant Vera Debin also accentuated their wealth:

> Their parents were wealthy, they had the best clothes, the trendiest clothes... the wealthy kids hung out with wealthy kids... the preps.

Cherry Gate stressed their upper-class status and visibility, and underscored, as Bill had, their tight bonding:

> The popular group, I guess we would consider like the upper, upper class... I would say their appearances and everything would make them bond together.

Recollections of two alumnae, Cecilia Hood and Betty Ames, graduates of 1996 and 1999, respectively, suggested that during their high school years, they had been the preps that Bill, Vera, Cherry, and others talked about. While neither Cecilia nor Betty identified themselves directly as preps, although Cecilia identified herself as being part of the popular group, both alumnae were of upper-middle-class background; fondly remembered their high school years as a time when they were involved in myriad sports, extracurricular activities, and leadership roles; and reported enjoying long-term friendships with peers whom they had known since early childhood. Betty explained:

> I played two sports every year, I was class treasurer, I did meetings for that, I was also in SADD [Students Against Drunk Driving]...I was in GT and AP honors classes...I was in school every day from eight in the morning and I didn't get home until 5:30...The class officers, we put a lot of effort into organizing [school events].[55]

While Betty and Cecilia held the same high profiles as their counterparts of earlier generations, they did not enjoy the same levels of participation in their organized social events. The worlds created by the preps, which had been patronized by the Divided Generation and tolerated by the Border-Crossing generation, were being boycotted by many non-preps of the Redivided Generation.

Over the three time periods studied here, there is evidence to suggest that school social events, prom in particular, had been the playground for Miller High's white female preps. Proms were a symbolic space within which, across time, the preps exerted progressively greater control over all aspects, even to the point of rigging the results of school-wide elections for the prom song to impose their own sentiments against majority vote, as Sophie Baker remembered about her graduating prom of 1985. The preps continued also to represent the school through yearbooks, without necessarily representing all students. Vera Debin explained how biased the yearbooks they created were:

> I think they [the preps] went around and tried to get pictures of all the clubs, and all the band, but it was pretty much them and their friends, all extra pictures.

In her testimony, Vera suggested that yearbooks would "probably be more interesting if people were assigned to a committee to do the yearbook, randomly selected. Across the three generations of Miller High, recollections suggested that preps image-managed their school.

However, while Linda Moss, a 1969 graduate, who remembered herself as one of "the doers...the people that were on the teams...putting out the yearbook...class officers," enjoyed a "big school spirit" and great participation in all school events that she helped organize, Betty Ames, thirty years later, felt frustrated by the lack of student participation. She remembered:

> The class officers, we put a lot of effort into organizing [school events.] It'd be frustrating when people didn't participate.... School spirit was terrible.

Preps continued to organize the visible social life of Miller High, but over time, fewer and fewer people showed up, and there is reason to believe that the prom became the event that non-preps could boycott, symbolically rejecting through nonattendance the world of the prep, the one peer identity primarily defined by "upper-class" socioeconomic status across generations.

When Sue Cohen remembered not caring about going to the prom, she also might have been, as a religious youth who distanced herself from excessive consumption, rejecting the ostentatious nature of proms. Over time, as yearbook pictures revealed, proms had become progressively more elaborate productions. By the 1990s at Miller High, they had evolved into showy productions that involved chartered limousines and expensive attire.[56] Russian immigrant alumna Vera Debin was among those who barely attended her prom. She peeked into the ballroom and almost immediately left with her boyfriend. She felt that it was merely a congregation of the prom organizers to show off their expensive clothes. She recalled:

> I wanted to see with my own eyes, walked in, saw what it was about, took pictures and left.
> *What was it about?*
> The[prep] cliques again, everybody looking at each other's dresses.

Preps' growing monopoly over time over the prom phenomenon might have contributed to institutionalizing the prom as a prep phenomenon. The institutionalized nature of the dance, inherited from the turn of the twentieth century,[57] would have been difficult to ignore. Yet by the 1990s, youth with strong religious convictions, some Russian immigrants and some "farm-country girl[s]" might have found it difficult to participate in a primarily prep-devised extravaganza, a ball for the princesses where king and queen are crowned. Furthermore, the general lack of school spirit throughout the 1990s and the many student nuclei, where "people kind of stayed in their groups"[58] and seldom crossed over between groups, might have made school organized parties by the few less appealing.

Although by the 1990s, as recollections of non-preps as well as preps suggested, the preps had become an elite group more unto themselves than at any other time since the Divided Generation, they were no longer only white. Black and white preps lived parallel lives, and occasionally intermingled. Bill Jackman and Vera Debin explained:

> Bill Jackman: "There were two popular groups. The popular black kids and the popular white kids. They didn't look down on each other, they would still interact and stuff. There was definitely two sides."
> Vera Debin: "A lot of the popular kids among the black kids would be friends with the [white] preps."

African American young women continued to fill the ranks of cheerleaders since the Border-Crossing Generation, and by the 1990s were also homecoming queens.[59]

While race, nationality, culture, and academic tracks primarily defined student relationships for blacks, whites, and Russians of the Redivided Generation, a segment of students across race, ethnicity, and culture congregated as youth of means who held high profiles. Thus, while African American youth of means Nat Right, a 1960s graduate, would be perceived by his white peers in the general track as "poor," by the 1990s, economically well-to-do African American youth were perceived as such by students across tracks. The prep was an identity primarily wrought around one's socioeconomic status. For preps, class trumped race and ethnicity. Russian immigrant Vera Debin recalled one Russian peer "who came here when he was a baby, his parents were very well off, he pretty much hung out with them [the preps], he hardly spoke Russian." As a Russian then, the less immigrant one was, the more money one had, the easier one became a prep. Under the umbrella of upper-middle-class status, the categories of race, nationality, ethnicity, and religion merged at Miller High. If one were wealthy enough, one could be a black, Russian, Jewish, or Christian prep, and more easily rule abiding, since one's personal life would be deeply interwoven with myriad school-sanctioned activities that one would participate in developing and nurturing.

When asked to reflect on their peers in general, preps Betty and Cecilia recalled, as had other graduates of the Redivided Generation, how students segregated into groups:

> Betty: "Everyone definitely segregated themselves...in the cafeteria...two tables of black kids, and then the Russian kids, the athletes...definitely students grouped within their cultural backgrounds."
>
> Cecilia: "People kind of stayed in their groups...The black kids did sit together. We started having a large population of Russian kids."

They also recalled accounts of racial fights, which they, unlike their peers in the general track, never saw close up. Commenting on an article that she had read in the *Community Times* about racial fights at Miller High, Betty remembered: "I never noticed [racial tension]. It could be that I was removed from anyone involved in it." Cecilia, as earlier discussed, remembered the expulsions of students who were transferred to another high school because of their involvement in racial fights, but she had never witnessed them firsthand. Preps Betty and Cecilia lived their high school years alongside incidents of violence that did not directly affect them. Rule-abiding students, they engaged in leadership positions and sports, and lived lives parallel to those whom some alumni remembered as "troublemakers."[60]

Just as students of lesser means remembered the preps and how they dressed, Cecilia and Betty also remembered students with lesser means, and how *they* dressed:

> Cecilia: "There were a couple of kids that didn't have a lot of money. They wouldn't dress so great."
> Betty: "Boys that were involved in the fire department, they wore flannel shirts and jeans—country boys."

While the 1970s generation paid little attention to clothes, in the 1990s, just as in the 1950s and 1960s, clothes and cars played captivating roles in the imagination of many students. Speaking from the prep's point of view, Betty recalled: "A lot of that behavior [people making fun of people] came from the cars people drove. People turned fifteen and their parents would get a brand-new Explorer."

Betty and Cecilia, just like the "upper class" and "doers" of the Divided Generation and the preps of the Border-Crossing Generation, "always expected to go to college," whether they invested in the grades or not.[61] Betty's recollections in particular revealed her loyalty to Miller High, a school that, like the preps before her, she represented to the world and to the school's students through yearbooks; a school in which she invested many extracurricular hours on social events to which too few students came. Speaking of her alma mater, Betty said:

> I can't imagine my life without it...I definitely felt that Miller High was a great school...I felt like I got a good education, probably better than at a private school.

While remembering the preps first and foremost as wealthy and self-centered, Russian alumna Vera Debin, with a sense of concern for fairness, said:

> They are the people with the school spirit. I can see why they are representing the school.

Black Youth

Although some wealthier African Americans counted among the preps and intermingled as preps, most African American students segregated as black students. David Randle, in commenting about the 1990s in Miller Town, drew a distinct line between African Americans who had grown up in the town and those who hadn't, and between his generation and the new generation of young blacks impatient with the white status quo.

> You've got to look at our situation. We grew up in the county. We kind of knew how things worked. We kind of knew what to do and what not to do. And then you've got twice as many people. I mean, what happened to the black population at Miller then [in the 1990s]? It more than doubled. So you're talking about a whole lot of people. I mean they came from the outskirts, they came from the city, and they didn't want things to change, or they wanted things to change faster.

When I asked David if I could interview his daughter, who was finishing up high school, to get an African American high school student's perspective of the latter part of the 1990s, he said that she would not want to be interviewed by me. The same occurred when I asked another African American Miller Town resident if I could interview her much-younger brother, who had attended Miller High in the 1990s. The implications of my whiteness within a U.S. reality, in conjunction with my PhD candidacy in education at the time of the research and the power differential inherent in the production of research interviews created in the eyes of two potential African American narrators, both graduates of the 1990s, a person to whom they would not want to grant an interview. They explained to me that their daughter and brother, who are "militant" about race issues, would not want to be interviewed by a white researcher. The agency of these two alumni cannot be ignored. They refused to give their insights and interpretations to someone whom they understood to represent an educational establishment that has failed black students. Their political statement reconstructed the power differential on their behalf. Their refusal to tell of their experiences is a historic act that must be recorded as clearly as remembered testimonies. It could have been omitted or, perhaps worse in this case, footnoted. Their refusal is part of the metadata that deepen the history of Miller High student relationships, but inconveniently interrupt the narrative.[62]

David's daughter's generation approached relationships with whites suspiciously, skeptical about the realization of racial equality in the United States. Recalling his experience as a coach at Crescent High, a county high school predominantly attended by black students, David recalled when their football team played Miller High in the late 1990s. His recollections revealed a racist principal.

> We had a game at Miller High. I'll tell you the name, name is Thomas Lawrence, the principal. I'm standing on the sideline with the principal of

Miller High. Crescent High is playing, and there's a penalty, and the first thing he says is. "Oh, here *they* go again" [David's emphasis]. I mean, how am I supposed to feel...I knew exactly what he meant: "They're all black so something is going to happen; they're getting ready to start something."

By contrast, the recollections of white "standard student" Heather Korn, a 1993 graduate, suggested the opposite—an administration that privileged black students to the exclusion of white students.

> One day they had a speaker come. All the black students were invited to go to listen to the speaker. None of the white students were allowed to go. They [the black students] were taken out of class. It was a big deal because, at the time, they [the black kids] were having a big fit, [it was] not too far from the Rodney King beatings. So a lot of people started saying if we only had something for the girls, or only for the whites, there would be a big problem. We're just sitting back and watching this happen.

Heather's either purposeful or slip-of-the-tongue use of "they were having a fit" as she referred to black students' feelings about the broadly televised cruel beating of Rodney King betrayed racial animosity on her part. Heather's statement, "If we only had something for the girls, or only for the whites, there would be a big problem. We're just sitting back and watching this happen," captured both the late twentieth century constructed concept of reverse racial discrimination, as well as the extent to which racism lived close to the surface in the 1990s at Miller High, no longer policed or buried deep as it had been during the Border-Crossing Generation. Since she suggested reverse racial discrimination, perhaps Heather did not know or ignored the fact that the Bible Study Group, which met daily, was attended only by white students who represented, as Vera described, "the ultra-religious right."[63]

Interracial Athletes

While black students mostly sat together, it was on the sports fields that students worked as teams and forged friendships across racial and ethnic backgrounds, as it had been since the early days of desegregation. Betty Ames recalled, "[A] lot of the basketball players were black, and the white basketball players would hang out with them." Still, it was only within certain sports that black and white students met in the 1990s. Throughout the 1990s the football team fared poorly,[64] and more attention went to basketball, soccer, and lacrosse. Black male students played neither soccer nor lacrosse,[65] so basketball was the only sport other than football where they played alongside whites. Throughout the 1990s, football lost the prestige it had enjoyed during the Border-Crossing Generation. Yearbooks, however, revealed black female students on soccer teams, as well as on softball and field hockey teams. Cecilia, who played soccer and lacrosse, remembered that her group of female jocks included "a group of black friends and two friends who were Indian."

White Youth Subcultures in the Midst of Racial, Ethnic, National, Class, and Religious Segregations

Peer groups of the Redivided Generation, unlike those of the Border-Crossing Generation, were white, and were local manifestations of broader international youth subcultures. Indirectly, through some graduates' memories, the punks and goths came into view. Vera Debin remembered the punks as defiantly anti-prep (the one group primarily identified by its superior buying power and upper-class status): "A punk would not go and talk to a prep, because he doesn't need their acceptance." In contrast to Vera's recollection, Bill Jackman remembered how one punk in particular crossed group boundaries:

> He was a punk kid; he was in a punk band, traveled around the country. He had a Mohawk. He was the most extreme-looking kid out of the whole crowd. He crossed over in all the groups. Every teacher loved him. He was a great kid. You look at him, you run away from him.

These were suburban youth whose affiliations with the punk subculture might have been more aesthetic expression than genuine economic and political protest against forms of commodity.[66] Still, Vera's recollection that a punk did not need the preps' acceptance underscores a more broadly shared reaction to consumerism and conspicuous consumption by a growing segment of Miller High students. In this sense, Vera's remembered punk shared with Sue Cohen of the Jewish Youth Group, and with Bill Jackman of the Fellowship of Christian Athletes and Christian Young People Club, an aversion of and challenge to consumerism. Of interest too is that some punks visited the Christian Young People Club, as Bill remembered: "We had more than a few punks [attend the Christian Young People Club]." Because the Christian Young People Club tended to be frequented by youth of politically conservative backgrounds, and in view of the one punk whom everyone loved, it seems safe to assume that Miller High punks espoused an aesthetic rather than political expression of the broader punk subculture. Thompson describes the suburban punk as follows:

> At the end of the continuum is the suburban youth who works for his spending money at the Hot Topic store in a mall, wears the store's apparel, and sells major label–produced CDs by Blink 182 and the White Stripes. This youth's cohort includes everyone who encounters punk primarily through videos on MTV or VH1. At the other end of the spectrum are the 'genuine' punks...who listen solely to independently produced anarcho-punk, steal or panhandle for their means of subsistence...and eschew most forms of commodity.[67]

Along with the punks, goths, who are generally nonpolitical,[68] constituted youth subcultures that were white, middle-class, and suburban. These were consumers of already commoditized pop cultures, even as some of the punks rejected the prep status. One seriously wonders whether school authorities or peers would have as easily accepted an Afro-punk.[69]

Whether rule-abiding punks, preps, or immigrant Russian Jews, or whether "down-home farm girls" or religious youth, these young people lived lives that ran parallel to those known as "troublemakers." Troublemakers, as Betty and Vera described, were targeted by the administration:

> Betty: "They [the administration] knew who the good kids were and the kids that got into trouble. They [the administration] were looking for the kids that got into trouble all the time."
>
> Vera: "The troublemakers would get treated differently."

Troublemakers

Some rule-abiding Miller High graduates of the end of the twentieth century identified troublemakers in generational terms. Roberta Jones, who graduated in 1991, recalled:

> I just remember the last couple of years in high school the younger kids seemed to be a little bit more out of control...I didn't think they were as eager to be friendly with teachers, they were more trying to get into trouble...I think that by the time I was a senior, I think the freshmen coming in, it just seemed like they didn't care as much. Yeah, back-talking—"I don't have to do it,' 'I don't care if you send me to the principal."

Drug users and drug dealers featured prominently as rule breakers in the memories of rule abiders of the Redivided Generation. Under zero-tolerance policies and organized efforts to create "drug-free zones," punishment for these rule breakers was swift. Cherry Gate recalled: "We always made sure it was a drug-free zone. If you got caught you were expelled and off the school property."

As the decade moved along, troublemakers took on more distinct identities and emerged overwhelmingly as Russians, blacks, and those referred to as "rednecks." Troublemakers were "standard-class" attendees. Reporting on Russian and black animosities, Ivan Strasky explained:

> Russian people were not tolerant of African Americans. Mostly Russians here are Jews, who came over because of discrimination. You'd think they'd be open-minded... endless cycle of discrimination... the most clashes came between the Russian community and African American.

Still, as earlier mentioned, recollections also indicated that clashes occurred between black students and the farmers. But some farmers, called "rednecks," also fought with Russian students and egged them on by saying, "Russian, go home." Ivan remembered:

> What has happened is at Miller High there is a distinct community of students who would be called, quote unquote, "rednecks." That is just what people generally say, people who are close-minded, hate Jews, hate blacks, could be categorized as KKK types. So that subpopulation at Miller High would threaten, by just being there, the Russians, because of their Judaism, and because of their being Russian. It's all kind of a broad assault on Russians.... On occasion I have heard slurs about Jews directed at Russian students.

There is reason to believe also that while clashes between "rednecks" and blacks, and "rednecks" and Russians, occurred primarily because of racist attitudes, they might also have occurred for economic reasons around drug dealing. Some graduates' recollections suggested that in the perception of Miller Town old-timers, the arrival of city blacks increased drug-overdosing incidents among black and white students as well as drug deals perpetrated partially by black "newcomers." Drug dealing, however, as Ivan's testimony suggested, was also taken up by many Russian youth who fancied themselves "Russian Mafiosos," wore "their leather jacket, stern look, a cigarette, and a defensive approach to things." Ivan recalled:

> I have heard stories about Mafia this and Mafia that. Kids pick up on this idea of Russian Mafioso... Russian kids, they are smart, Unfortunately their business was drugs.

It is reasonable to assume that Russian and African American drug dealers might have perceived each other as enemies when competing for the same market, especially in view of their loyalties to their respective racial and ethnic groups. Ivan said, "There is a very strong cohesion in the Russian community...Russian culture is very loyal."

Beyond cultural loyalties, many of the Russian troublemakers and "troublemaker black kids" shared two other characteristics that might have contributed, ironically, to their animosity toward each other. Students of both groups tended to come from the city:

> A lot of the Russians moved from the cities to suburbia, which is a big shock. In the city, it is a tight community. You walk everywhere, go to café, bar, everything was city oriented, Then, hit suburbia and oh, what do we do![70]

This youth had been transported from the high-energy life of the city to the comparatively lackluster life of suburbia. While Miller Town was suburbia in the midst of accelerated urbanization, it lacked sites of action that a city proper offered: "What do we do!" One can imagine that these young people itched to make something happen in their immediate surroundings. Proud,[71] loyal, and restless, they also shared a fundamental mistrust of the American educational system. Russian youth had a "lack of confidence, lack of faith in the American system." Ivan explained:

> They were that pole of the Russian community. Nothing extracurricular; they were not committed to going to college, or if they were, they had a negative perception of school, that they didn't want to have anything to do with school...
>
> There is a strong anti-American feeling...what can an American teacher authority teach me?

Russian "troublemakers" could often be seen arguing with their teachers, and teachers would anticipate their challenges:

> From the way I know teachers and the way I talked to them after the years I left, there is this perception that teachers definitely know the difference between a Russian kid and average American Joe. I don't think teachers ever treated kids differently willingly...but when students started arguing with the teacher, the teacher would be more likely to butt heads with the Russians...just because it was almost expected, because us Russian kids would always argue pretty well, I don't know if that is Judaism or the Russian culture...as soon as the teacher sees she is losing an argument, she uses her authority. There is a predisposition against Russian kids, because that general stereotype.[72]

Similarly, some African American students of the Redivided Generation might have felt that black people had been overstudied by representatives of the white race, who in turn had too little to show for their research in terms of advances in the education of black youth in the United States.[73] Pride and loyalty to their respective racial/ethnic cultures, city restlessness, and fundamental mistrust of the American educational system could have made the black and Russian troublemaker students allies in rebelling against "the system," but it didn't. Racism on the part of some of the Russian youth, and perhaps competition in the drug market, made rivals out of these rule breakers. Furthermore, young Russian rule breakers seemed to revel in competing to break rules first, and to have things "happen sooner to [them]":

> First to get drunk, the first to [smoke] cigarettes, the first to use marijuana, the first to use ecstasy...huge use and abuse of ecstasy, a lot of people fell into that, that always happened earlier in the Russian community.[74]

There was also the question of defending one's identity:

> I don't think the African American community ever started anything. It was defensive, on the African-American part, just as it was on the Russian part, just as it would be in any minority. Defend their identity.[75]

Male rivalries had dominated the memories of previous generations of graduates, but by the 1990s, recollections emerged of girls fighting.[76] Black and white girls fought their own racial battles. Heather Korn recalled:

> It was a black girl and a little white girl who was very snotty, she made it known that she was very racist. She was in the cafeteria line, she looked at this girl funny, and the black girl got ticked off because the white girl looked at her funny. Somewhere in the middle of everything, while we were eating lunch, you heard somebody scream, then you heard loud bangs. The [black] girl came up behind her, grabbed her hair, and slammed her face into the cafeteria table. They both ended up getting suspended, so it wouldn't look like a racial issue.

Girls could be seen "scratching"[77] while fighting over boys. Roberta described: "There was one fight that was outside of the lunchroom...It was two girls, probably fighting over a boy." Ivan described Russian girls who might count among the troublemakers and rule breakers as: "Strong Russian bitch. Don't mess with that girl, she will talk you off." By the end of the 1990s, males and females equally engaged in physical fights and verbal aggression.

Exacerbated Segregation

By the end of the twentieth century, Miller High students segregated not only along racial, class, religious, and ethnic/national categories of identity, but also across those categories, as rule abiders and rule breakers. To complicate the matrix of segregation, students were further divided in their experiences of each other and school authorities along "standard classes" and "upper tracks," as they had been in the 1950s and 1960s. In the following chapter I discuss the self-segregating tendencies of students of the Redivided Generation in light of structural forces, and I examine more closely the clashes in student ideology between youth who espouse consumerism and those who oppose it while remaining overall a politically inactive student body.

Chapter 6

Oppositional Self-Segregation: A Student Body Sensitized to Discrimination (1986–2000)

For the Redivided Generation, class and race—divides to which this generation also added those of religion, ethnicity, and nationality—were no longer categories of identity by default, as they had been for the Border-Crossing Generation, but rather public statements of allegiance to particular social interests and world views. This was a purposeful generation not so much interested in "getting along" across categories of social identity as was the Border-Crossing Generation, but more of asserting differences across social categories. Being African American, a Russian immigrant, Christian, or Jew; working class, middle class, or preppy and rich were identities embraced proactively and even defiantly in opposition to other identities. Moreover, the particular geography and demographics of Miller Town by the end of the twentieth century, now an urban suburban town with leftover rural-suburban roots, further divided residents into old-timers and newcomers, divided African American residents along generational lines, and divided whites along national lines.

These divisions along race, class, nationality, ethnicity, religion, and generation, as discussed in the previous chapter, were not uniformly clear-cut. That is, significant wealth trumped racial and ethnic divisions, since well-to-do black, white, and Russian youth socialized as preps. As had been the case for students of the Divided and Border-Crossing generations, athletes crossed racial lines on sports fields, as did students attending the prestigious upper academic tracks. Preps, athletes, and upper-track students notwithstanding, many more students of the Redivided Generation segregated within standard classes, where they also engaged in racial fights. This was a generation that adopted the broader social categories that define census data for new purposes—for personal defense and empowerment, albeit not broader political change.

Peer Relationships

Balkanized Student Groups

Miller High students' self-segregation of the late 1980s and 1990s resurrected stereotyping, something that had been fought against by the Border-Crossing Generation and suffered by the Divided Generation. Russian immigrant Vera Debin recalled:

> I remember some incident, somebody got into a fight, with one of the [Russian] guys in our group, and then his mother got a restraining order against the Russian boy because she was afraid that the rest of the Russian kids would come to her house and beat her boy up.[1]

Russian youth were more likely to be stereotyped as aggressive or even criminal; black youth as drug users; white farm youth as rednecks; and religious youth groups, who were overwhelmingly white, as ultra-right-wing conservative. But these stereotypes were renegotiated by the students in ways that defined their loyalties to particular group identities within Miller High and the broader community, generating myriad combinations of oppositions. Recollections, as discussed earlier, revealed some of these identifications in contrast: a native-born Jewish religious youth defined in contrast to the stereotype of the Jewish American Princess; a Christian religious jock defined in contrast to overindulgent, alcohol-drinking jocks; a Russian immigrant youth defined in contrast to an "American" youth; a troublemaker immigrant Russian youth defined in contrast to an African American youth; a white farm youth defined in contrast to black, Jewish, and Russian youth; and segments of all youth of lesser means in contrast to the preps.

The particular demographics that both significantly increased and diversified the population of Miller Town and its high school, the stringent disciplinary technologies delivered by distant school authorities, the ever-escalating tracking,[2] the broader intensification of the credential race and the undemocratic power of the preps all combined to create an environment conducive to the balkanization of student groups and to students' disengagement from school life.

Explaining the Balkanization: The Intersecting Roles of Demographics and School Structure

The sudden and overwhelming influx of ex-city dwellers and immigrants to Miller High drastically increased the number of students who were not only new to the neighborhood, but who also had not practiced peer relations within desegregated schools, because they came from myriad schools beyond Miller High feeder schools—from the city and from Russian schools. Youth of the Redivided Generation were doubly strangers: to each other and to neighborhood kids attending Miller High. The school was also doubly strange to them: directly, because they did not grow up attending the town's fall festivals organized on its premises or the games and presentations in its gym and auditorium, and indirectly, because they did not have older siblings and relatives to relay school traditions or gossip about teachers. Not surprising, as newly arrived outsiders within an overcrowded Miller High student population, these young people sought the company of students with similar backgrounds. As Russian immigrant graduates put it:

"It was easier to speak Russian."[3]
"I was drawn to the Russian crowd...hanging out with Russian friends, dressing the Russian way...speaking Russian."[4]

Alumni testimonies across time overwhelmingly revealed that when entering Miller High without prior or subsequent involvement in extracurricular activities, students of similar racial, class, and ethnic backgrounds naturally tended to congregate.[5] This was particularly the case for Miller High students in the late 1980s and 1990s. Amid the overcrowding, they huddled in groups for comfort: the Russian immigrant students for whom "America" was still foreign, and the religious students for whom mainstream U.S. culture was too lax in its mores. Other students huddled together out of comfort: the preps who simply knew each other forever and traveled the same upper-class circles; the African American students, newcomers and new generations, who no longer adjusted their attitudes to fit the white world of Miller Town as might have the previous generation, who "knew what to do and what not to do"; and the "farmers" or "rednecks" who felt besieged by ex-Communists, Jews, and blacks all at once and wanted people to "go home," as recollections of others impressionistically sketched out.

Graduates across time periods consistently reported that students associated with certain groups and not others out of a need for shared comfort, shared upbringings, shared language and musical tastes—that is, they congregated for cultural reasons. In the 1990s in particular, in the context of a seriously overcrowded school and an institutional environment that exposed its students by stripping them of privacy to the point where students' personal effects could be searched if administrators found them suspect, finding comfort was difficult. Comfort became the other body whose language and cultural traditions reminded one of home. Youth group associations wrought around familiarity were the most natural and spontaneous expression of the human need for shelter and comfort in the face of institutional discomfort created by overcrowding, intolerant and strict punitive disciplinary strategies, and an influx of foreigners.

However, for a segment of African American and white youth of the Divided and Border-Crossing generations, the comfort of shared culture had always been inextricably linked to race. While this fusion of race and culture was not set in stone for the earlier generations, it became more rigid for the Redivided Generation. Being a black Miller High student definitely meant being of a different culture than a white Miller High student. Many young county blacks at the end of the century submerged their own particular upbringings under the racial category of "black," and identified with ex-city black youth—often to the dismay of the earlier generation of long-time black county residents, who lamented losing their young to drugs and the bad influences they attributed to city blacks. One African American mother, a long-time county resident, said, "I don't want them blacks from the city coming into my town ruining Miller Town. That's how old blacks in this area feel." Many white students stereotyped as "rednecks" also conflated race and culture: Being white American meant being anti-black, anti-Russian, and anti-Jewish. For many white Russian immigrant students, being Russian meant being Jewish, even though they did not practice the Jewish religion—immigrant Russian youth identified with being of Jewish culture. Race, ethnicity, nationality, and religion, rather than shared affinities across those categories, became dominant markers of identity for the Redivided Generation.

Because student groups are inextricably linked to notions of individual identity, more research is needed to explore the influences of multicultural curricula on students' identity constructions over time. Toward the end of the previous century, it is more likely that native-born public school students were exposed to the histories and points of view of U.S. residents from different racial, class, and ethnic perspectives. Perhaps graduates of the Redivided Generation were schooled in dividing the world according to categories of race, class, ethnicity, gender, and so on; and schooled in seeing each other belonging to one of those groups. Perhaps the self-segregating tendencies of the Redivided Generation reflect institutional influences more than demographic effects.[6] Still, school policies and institutional norms exacerbated the nefarious effects of overcrowding and the disorienting effects of accelerated diversification at Miller High during the last decade of the twentieth century.

There is reason to believe that student loyalty to racial and national groups were intensified by the frequency and severity of punishment for transgressions. Vera Debin and Ivan Strasky remembered how they stood up for their Russian peers, even when they risked getting in trouble themselves:

> Vera Debin: "Somebody said something wrong, or pushed you out of the way... If somebody would say something in front of a Russian student, the other Russian students would kind of back him up. So, it would be scary."
>
> Ivan Strasky: "There have been a lot of times—'Oh, that black kid said something about the Russian kid'—and then, 'You guys, after school, parking lot.' Countless times, and I'd go, because in the back of my mind, 'Ivan, what are you going to do? You're not going to fight, because if you get in trouble, it's not worth it. But if something bad does happen, you have to stand up.' Even in my academic mind, when I knew what I had to do to succeed and go to college, I still had this association... that I have to go."

Judging by Ivan's testimony, one can easily extend the same kind of loyalty to the black students who engaged in the parking lot fights that Ivan described, as well as to the "redneck" students with KKK tendencies. While seeking comfort might have primarily propelled ex-city blacks and Russians to retreat into their own groups of black and Russian students, zero-tolerance policies might have contributed to sealing and consolidating group loyalties around racial and ethnic boundaries. Loyalty to racial or ethnic groups might further explain why young county blacks would more easily connect with city blacks, and why the mother of a white boy assaulted by a Russian boy might have assumed, as earlier mentioned, that "the rest of the Russian kids would come to her house and beat her boy up" out of loyalty to their Russian peer.

Rigid school policies of the 1990s contrasted dramatically with those of the Border-Crossing Generation. In the 1970s and early 1980s, when students knew each other from attending the same elementary schools, when the population had remained fairly stable, and when school policies had been relatively lax in comparison to zero-tolerance policies, students experienced greater interracial and cross-class friendships and greater intermingling across categories of race and class. During the 1970s, the Border-Crossing Generation also basked in the broader lingering effects of the civil rights movements. But by the 1980s, as Tyack and Cuban suggest, "Policy talk about schools stressed a struggle for national survival and international competition...In such periods, policy elites want to challenge the talented, stress the academic basics and press for greater coherence and discipline in education." The authors remark that by the close of the twentieth century, in the "late 80s and 90s...conservatives and liberals alike...called for national standards."[7]

The contested institutionalized nature of the preps, the one group of students who continued to construct the public image of the school across time periods, might have further contributed to lessening student encounters across groups. While stories of alumni who graduated in the 1950s suggested that students collaborated with preps and participated in school social events across class divides—even across race divides in the later years of desegregation in the 1960s—and while the stories from the Border-Crossing Generation suggested a tolerated prep contingency and a continued participation in prep-organized events, albeit not without criticism, stories from the Redivided Generation revealed outright rejection of preps through prom boycotts and moral indictments of their values, as earlier discussed. The ironic and unintended consequence of student boycotts of prep-orchestrated social events might have been that they also robbed students of opportunities to meet across social divides. Neither recollections nor yearbook data suggest that alternate groups of students might have taken charge of creating social gatherings parallel to those of the preps. In fact, it was Vera Debin, a critic of the preps, who underscored that "[t]hey are the people with the school spirit. I can see why they are representing the school." Needless to say, it would have been easier for native-born youth whose white middle- and upper-middle-class backgrounds aligned with school authorities to feel most empowered at Miller High and least challenged or suspected by teachers and administrators.

The Class Issue

The upper-class economic status of white, black, and Russian preps would have plausibly further galvanized the frustrations of those students identified as "farmers" or "rednecks," as they felt snubbed by preps, who by the end of the century included black students. The reader will recall lower-class white students' refusal to recognize Nat Right's middle-class status in the 1960s because he was black. The reader will also recall Cherry's disdain, in the 1990s, for the preps, Cherry who identified herself as "a back-home farm-country girl." These, among other testimonies earlier discussed, further suggest that the undercurrent of animosity expressed by some of those white students identified as "farmers" and "rednecks" toward the visibly rich and toward blacks, an undercurrent that might have been repressed in the 1970s and early 1980s by vigilant white peers, surfaced and exploded in the 1990s.

The preps of the 1990s came to represent a coalition of races and ethnicities bound by superior consumer power. For the "rednecks" of the community, who witnessed the disappearance of the farms they had worked on for generations as the general consumer lifestyles of the city took over, this coalition might have seemed overwhelming, and more than that, as recollections seem to suggest, was an affront to their "farm country," "down-home" way of life.

Finally, that graduates of the Redivided Generation were remembered as rule abiders or rule breakers suggests an environment more akin to that of inmates in a prison than of students in a school. Remembrances of an explosive cafeteria where racial fights erupted and where racial and ethnic threats and slurs were exchanged describe the atmosphere of a prison mess hall rather than that of a place for students to congregate in safety and to eat. There is little doubt that school authorities were taken aback, in the late 1980s and 1990s, with the sudden influx of youth, both immigrant and from the city, as they scrambled to build an annex. With the privilege of hindsight, however, this historical analysis shows that segregating forces had already been entrenched in the organizational structure of Miller High, which fully came to view with the avalanche of newcomers. Structurally inherited tracking divisions continued to isolate students into parallel worlds[8] and to expose students in lower tracks to violence more readily than students in the upper tracks. A cafeteria devoid of teachers' and administrators' involvement and role modeling allowed for racial and ethnic hatreds to erupt.

Youth and Authority

At the end of the 1990s, the highly competitive climate for access to colleges increased student rivalry in the upper tracks. As a result, student-to-student relationships hardened both within and across tracks. Achievers competed against each other, sometimes ferociously, as Sue

Cohen's testimony regarding the aspiring valedictorian revealed, and as suggested by the recollections of Bill Jackman, Ivan Strasky, and Vera Debin. In the 1950s and 1960s, white students easily coasted through upper tracks in large part because the upper tracks were economically rather than academically defined, and were by and large reserved for those who could afford college tuition. Those most likely to work hard in the upper academic tracks were black youth. In the 1970s and early 1980s, students across race and class invested minimal effort in school, particularly because the older teachers held low standards for their charges. In general, students of the 1970s and early 1980s were less invested in studies or in pleasing teachers. By contrast, the Miller High upper-track student in the Redivided Generation felt the pressures of academic competition as more middle-class youth competed for choice colleges, and as credentialism created expectations for universal college attendance. Of the previous generations' students, perhaps the experiences of 1950s graduate Norman Good, the African American alumnus who attended the academic track where he had to prove his competitive spirit, might have approximated those of the 1990s academic-track students. Still, as Norman Good recalled, he "competed in a good-spirited way for grades." Memories of 1990s graduates revealed how rather than "good-spirited," competition was strategic and stressful.

Moreover, achievers constructed their upper-track status as proof of their superior qualifications by pointing out that students attending standard classes were either of lesser intelligence, as Ivan's recollections emphatically suggested, or not as hardworking, as Bill's and Vera's testimonies revealed. In this sense, graduates' recollections echoed those of the Divided Generation who constructed "lower-track" students as less deserving.

When Miller High students focused on ensuring their particular student status rather than their peer status, as was the case for both the Divided and Redivided generations, then disdain by upper-track attendees for lower-track attendees intensified. Within a dividing and hierarchical system, the more teacher-centered the students, the more likely they were to construct each other hierarchically, according to institutional divisions. The reader will recall how in the 1970s and early 1980s, teachers' divisions of student status according to tracks were not embraced by the Border-Crossing Generation, whose respect for teachers had plummeted when compared to that of the Divided Generation in the 1950s and 1960s.

What is of importance here is that teacher-centered students—students invested in either pleasing their teachers (1950s and 1960s), or befriending them as they sought competitive advantages (achievers of the late

1980s and 1990s), or not crossing them to ensure protection in a zero-tolerance climate (rule-abiding standard-class attendees of the late 1980s and 1990s)—integrated notions of one another as superior or inferior as defined by classes taken and grades acquired. Thus Ivan and those students institutionally recognized as belonging to that "pole" of Miller High youth whose academic status set them apart from the majority of students considered someone like Cherry, who attended the general classes, of lesser merit. Cherry, in turn, placed herself hierarchically lower than GT attendees Ivan, Sue, Bill, or Vera. Recall how Cherry, Heather, and Roberta situated their place in school somewhere between "the group that did drugs" and the "higher group... that you knew were really smart." Tellingly, as if in an attempt to avoid the lowest status along academic tracking, they placed themselves above those who "did drugs," a qualification that does not define one academically but socially. Also telling, not doing drugs accorded them a status of good conduct, which in turn ensured acceptance and belonging to an institution hypervigilant against transgressions and readily apt to expel.

However, while the Redivided Generation resembled the Divided Generation in that it was school authority centered rather than peer centered, students of the Redivided Generation did not develop authentic personal relationships with school authorities as did graduates of the 1950s and 1960s. Rule-abiding students of the Redivided Generation objectified or abstracted school authorities. Achievers saw them as a "means to an end." Recall Ivan's words: "Kids at that level always have great relationships with teachers because they do whatever it takes to get the teacher to like you... To get an A you need a relationship with the teacher." Standard-class attendees saw school authorities as ivory-tower keepers of the law whose presence was made manifest only through expulsions and punitive actions. One alumnus's recollection summed it up: "I cannot remember seeing the principal walk the hallways... You stayed away from the office."

Rule breakers also objectified school authorities and continued to clash with teachers, whom they constructed as "the system." However, unlike the rebels of the 1950s and 1960s, who could be saved from endless detentions by a shop teacher taking personal interest in their lives, rule breakers of the Redivided Generation were less likely to be pardoned and were also less likely to develop meaningful personal relationships with teachers. Meaningful relationships with teachers were also less likely for successful students and those fully integrated into the life of the school. GT student Vera Debin recalled: "There was a very few teachers that I had any kind of relationship with other than in class," and the prep jock Cecilia Hood said: "I don't remember the teachers that much." Graduates of the Redivided

Generation talked about teachers in abstracted terms, and none of those interviewed spontaneously recalled or named a particular teacher as had graduates of the 1950s and 1960s—further underscoring the segregation between teachers and students, and between administrators and Miller High youth by the end of the century.

The cafeteria continued to be a herding place where students were left to their own devices to establish relationships in isolation from their teachers, who ate in their own quarters. Students, segregated from their teachers during mealtime—a time notoriously reserved, across cultures, for community and communal relationships—further segregated from one another as they sought familiarity and comfort. The teacher-student segregation vividly came to light in Ivan's description of his privileged position with teachers in a school system where relationships are constructed hierarchically. When Ivan said, "I was president of my class, I'd always be in the lunchroom with the teachers," he made it clear that he was not part of the student cafeteria crowd, but that he had access to the private world of faculty who ate away from the students who were considered lower in the hierarchy of relationships.

It is little wonder then that within the space of the cafeteria, racial, ethnic, and class animosities flared up in the overcrowded Miller High of the 1990s. Carnoy and Levin's remark that "schools continue to provide Americans with a social experience that is markedly more egalitarian and more open to free choice and possibilities of self-realization than anything that is available to them in the realm of work"—a remark that might have captured to some extent the experiential realities of some students of the previous two generations—no longer applied to the Redivided Generation. By the end of the twentieth century, Miller High was a place where students participated less and less in the life of their school, and where competitions within upper and lower tracks were as stressful as those found in the world outside: Achievers competed more fiercely to keep ahead; rule-abiding standard-class students held their breath hoping not to be sucked into what they perceived to be growing violence around them; and rule breakers broke rules with greater consequence than ever before as they were apt to be judged in courts as adults.

A Student Body that Negotiated Discrimination but Remained Politically Inactive

Native-born white graduates Bill Jackman, a Christian conservative youth when he attended Miller High, and Heather Korn and Cherry Gate, rule-abiding conservative standard-class students, reframed discussions around discrimination also to highlight "reverse discrimination"—discrimination against whites by blacks. Jackman spoke of people in school "that were racist either way." Heather Korn complained about privileges accorded black youth that were denied to white youth.

Heather's recollection spoke of unfair treatment against whites and of their disempowerment as they were just "sitting back and watching this happen." The reader will also recall Cherry Gate's fear of black students as she recalled death threats against her sister implying black racism against whites at Miller High. Some white Russian youth also perceived that they were discriminated against by "American" schoolteachers just because they were Russian. But Ivan Strasky further suggested that many were quick to call discrimination in situations that in fact were teachers' standards for academic performance for all students:

> Whenever the teacher challenged them, not because of being Russian, but because the teacher was being a teacher—"Why didn't you do your homework? Why are you late?"—they would automatically be defensive, and often perceive it as discrimination.

Recollections suggested that segments of native conservative white youth and Russian immigrant youth negotiated the idea of discrimination to underscore their victimization by blacks or by the "American" educational system, thus suggesting a student body fragmented into communities that were hypervigilant of discrimination against their own groups' racial, ethnic, national identities. Defensiveness and defense of group identity along racial, ethnic/national divides characterized this generation. Of interest is that the Redivided Generation did not recall self-segregation on the part of students along gender divisions as had the previous two generations, or discrimination by school authorities along gender divisions as had the Divided Generation of the 1950s and 1960s—perhaps suggesting a more broadly shared sense of gender equality by the end of the century.

This hypervigilance and hyperdefensiveness, particularly along racial identities, comes even more clearly into view when compared to previous generations' racial relationships, and even more so when comparing young women's experiences in the lower tracks. In the 1950s and 1960s, racial border crossings between young women were timid. The reader will recall how African American alumna Doris Right, who graduated in 1958, having transferred to Miller High as a junior from the all-black school she had been attending prior to desegregation in 1956, described a white girl's tentative attempt at making friends.

Twenty-three years later, the description of racial border crossings between African American and white adolescent women would prove more daring, but also more dangerous for African American young women. The reader will recall 1981 graduate Theresa Randle's experience with her friend's KKK father.

By the mid 1990s, however, stories emerged of racial fights between young women.

Young women's relationships across race in lower tracks progressed at first from timid to daring between the Divided and the Border-Crossing Generation, then shifted into racial violence by the end of the century. Until the Redivided Generation, young women's interracial relationships had been more conciliatory than young men's. This reversal in trend among young women in lower tracks toward greater interracial relationships may be in great part attributable to institutional messages about race relations transmitted across time periods.

During the early years of desegregation (1956–1969), the institutional message regarding race relations was that black students should overlook racism and be patient. During the Border-Crossing period (1970–1985), the institutional message was that the efforts for inclusion should be made by black students, even as many youth crossed racial, gender, and class borders without much reference to school or parental authorities. The reader will again recall the testimony by African American alumnus David Randle, a 1976 graduate: "[N]obody approached us about [joining sports]...so what happened...a lot of kids they didn't get involved in anything."

During the Redivided period (1986–2000), the message was that the responsibility for violent racial altercations should be equally shared by white and black students, as Heather Korn's testimony suggested.

From "bear the racist insults" in the 1950s, to "find your own way into mainstream life" in the 1970s, to "assume equal responsibility for racial violence" in the 1990s, the institutional messages across generations consistently betrayed a "washing of the hands" by school authorities of any responsibility for helping youth integrate authentically across racial and other social divides—authentic integration understood as "intellectual and social engagement across racial and ethnic groups,"[9] as addressed in the Introduction. It is not surprising, then, in light of continued lack of adult positive involvement in the creation of an authentically integrated school climate, that youth redivided. Consistently, the onus for integration was placed directly on the shoulders of students. The students, in turn, by the end of the century, were ready for battle, whether along racial or nationalistic divides. From complaints about teachers' unfair treatment to actual racist peer behavior, students readily voiced accusations of discrimination and fought racial fights.

Evidence suggests that just as the two preceding generations, the Redivided Generation was not politically involved. The balkanization of students within an overcrowded school environment, groups of students with adversarial dispositions toward the U.S. educational system and the U.S. government more broadly, and a segment of the student population whose energies were consumed by commitments to scholastic success above all else might have contributed to a more narrow, immediate, and personal focus on part of students. Then, too, students who decried the consumer-oriented factions of Miller High and who might have found a basis for political action around issues of consumerism and class inequity were students whose group affiliations were oppositional: Russian, "redneck," conservative Christian, and punks expressed their affiliation aesthetically rather than politically.

* * *

The insular and xenophobic tendencies of "rednecks" and their long-standing racism; the mistrust and survival skills imported by newly transplanted black city youth, and by many Russian youth eager to change their fortune in "America," all the while begrudging it its Americanism; the zero-tolerance policies that divided those who toed the line and those susceptible to expulsions; the explosive overcrowding; and the "prep factor" combined to create strands of clashing identities claiming their space and legitimacy within the halls of Miller High, and within an educational system that by the end of the twentieth century was scrambling, yet again, to reclaim its relevance.[10] After forty-four years under *Brown v. Board of Education*, Miller High students segregated according to adult-devised census categories of identity, and lived parallel lives further divided from those of school authorities.

Conclusion

Authentic integration at Miller High over the second half of the twentieth century had been the experience of too few students across generations. Interracial friendships that lasted beyond school years were mostly forged during the 1970s and early 1980s, an anomaly in this fifty-year history. A propitious convergence of social forces sustained the Border-Crossing Generation: a broadly shared middle-class status across racial divides; a population of black Miller High students who were both minorities and mostly children of families long established in the region; a student body accustomed to desegregated institutional settings; and a national consciousness of civil rights that stirred the imagination of an otherwise conservative student population in a conservative town.

Structural Forces

One may further claim with some confidence that without the national backdrop of the civil rights movement, the Border-Crossing Generation would not have forged relationships across divides to the extent it had. Local conservative politics mitigated the influences of broader national forces. Blacks with roots in the community that extended back to the 1800s, whose small businesses received white patronage, and who were "familiar with getting along with white people," and whites in the broader community for whom the status quo was comfortable, together created a local context within which Miller High youth would not engage politically to demand curricular, extracurricular, or other changes that peers in other parts of the country were claiming. The Border-Crossing Generation democratized peer relationships despite institutional norms but while not challenging them. The result, however, was that a segment of students—disenfranchised lower-class white youth and marginalized black students—continued to segregate during the most democratic period in the history of Miller High. The social context that proved favorable for nurturing cross-racial friendships, in particular the broadly shared middle-class status,[1] and the socially emancipating effects of the civil rights movement on black and white students, might also have been the factors that kept poorer whites and culturally critical young blacks of Miller Town segregated—the poorer white students because they were economically marginalized in a predominantly middle-class world; and the critical black youth, because they were culturally marginalized in the overall conservative Miller Town of their times. In the midst of Miller High's most authentically integrated period, segregation of poorer whites and disengaged blacks continued, underscoring the perdurability of segregation among students in Miller High over time.

Across time periods, too, cross-class friendships within and across racial divides were also rare. For each generation, they occurred mostly in upper academic tracks and a few extracurricular activities, most notably sports fields. These spaces were the least populated and most insular—spaces for the chosen few. Poorer students' possibilities for participation in upper academic tracks or extracurricular activities, which proved central to developing relationships with peers across diverse backgrounds, were restricted. Those students of poorer backgrounds who could join the upper academic tracks or play in organized sports increased their chances of forging friendships across class as well as race divides. Within elite spaces, encounters across differences were more likely to be friendly. Within the more populous general and vocational tracks, on the other hand, they were more likely to be unfriendly.

Indeed, the general and vocational tracks remained overall divisive spaces and incubators of racial violence. More important, animosities were not only reproduced, but also intensified within these tracks. In the confined spaces of Miller High, students could not avoid each other by dispersing into alternate realities as they might have in the larger community. The cafeteria also remained a place where students continued to segregate as they sought comfort in the familiar, demonstrated loyalty to their own, and consistently were left by school authorities to fend for themselves. During the 1990s, when demographic shifts brought into school perimeters Russian immigrants and black city youth from segregated schools and mistrustful of the school system and of each other, spaces such as the cafeteria only exacerbated their segregating tendencies[2] as they sought refuge in familiar ethnic, racial, religious, and economic relationships.[3]

Across time periods, too, changes in student relationships and demographic configurations were accompanied by changes in disciplinary measures and types of teacher-student relationships. Although a clean, linear progression and one-to-one correspondence between disciplinary measures, demographics, and student relationships across generations cannot be established, given the evidence, it is reasonably safe to advance that overall, as the student population grew and diversified, school authorities became more distant and disciplinary measures became more severe. Evidence also suggests that zero-tolerance policies at the end of the century might have exacerbated student segregation by consolidating and hardening student loyalties along racial and national lines.

The case of Miller High undermines the assumption that desegregated comprehensive schools have actually lessened racial prejudice[4] since *Brown v. Board*. First, this history challenges Grant's affirmation of an achieved "genuine racial equality." By extending educational historian Gerald Grant's story of Hamilton High beyond the mid 1980s into the end of the twentieth century, this history reveals that Grant's affirmation holds only under certain demographic conditions. Similarly, this analysis complicates the findings of a recent groundbreaking mixed-methods educational study, *How Desegregation Changed Us: The Effects of Racially Mixed Schools on Students and Society,* which examined the experiences in several desegregated schools of graduates of the class of 1980.[5] The study findings suggest that "school desegregation fundamentally changed the people who lived through it...desegregation made the vast majority of the students who attended these schools less racially prejudiced and more comfortable around people of different backgrounds." The findings in *How Desegregation Changed Us* confirm some of the findings in this history: Those who attended high school in the 1970s and early 1980s (Border-Crossing Generation) more readily forged friendships across racial divides that lasted throughout their lives. However, this history has further revealed that for most students such experiences did not continue into the end of the twentieth century. More important, this analysis has shown that across time periods, students in the general and vocational tracks did not experience desegregation in positive ways. The authors of *How Desegregation Changed Us* interviewed people who had experienced desegregation twenty-five years ago, when the nation rode the crest of the civil rights movement, and interviewed "educators, advocates, and local policy makers," a population of people less likely to have attended the vocational and lower academic tracks.

Experiences

This analysis also shows that by the end of the twentieth century, Miller High no longer provided its students with experiences "markedly more egalitarian and more open to free choice and possibilities of self-realization than anything that [was] available to them in the realm of work," as Carnoy and Levin have suggested. While middle-class white women in the 1950s and 1960s, and middle-class black women of the 1970s would be shocked at the discrimination they encountered in the workplace upon graduation, evidence of their more egalitarian experiences within the confines of Miller High, students of the 1990s experienced Miller High *as* the workplace. By the end of the twentieth century, fewer students participated in the social life of their school than ever before in its history. Those attending the upper academic tracks competed more fiercely, and developed instrumental, professional-like relationships with teachers as they negotiated their way to the top. College attendance having had become de rigueur—these upper-track youth engaged in competitions almost as vicious as those found in the marketplace. On the other end, rule-abiding standard-class attendees welcomed security measures they felt protected them against what they perceived to be escalating violence around them; and rule breakers broke rules with greater consequence than ever before, expelled if need be for their offenses and sentenced as adults in courts.[6] While to a greater or lesser extent Miller High had always served as a holding place for many of its youth, in particular male youth, over the fifty years of this history, this role intensified with the unwitting consequence of increasing (rather than lessening) social unrest and violence among many students, even as it provided a few safe spaces for a select few to forge cross-racial and cross-ethnic relationships.

On an intimate and individual level, there are the darker and brighter stories that each generation told. The darker story of the Divided Generation of the 1950s and 1960s told of the burden of early desegregation borne by local black youth assaulted by white youth, and by young black women who were further subjected to sexual harassment by white males; of the humiliations of the very poor and of pregnant women across race; of the alienation of male students inept at studies; of the disproportionately privileged lives of white middle-class young women when compared to those of black middle-class young women, and of their favored status when compared to all other students. On the other hand, the brighter story of the Divided Generation told of young black and white women's impulses to cross over divides against entrenched racial mores of the early days of desegregation; and of black and white neighborhood friends having lunch together within the spaces of the cafeteria while blacks still purchased goods from the back doors of the community's white-owned stores; of young black men's academic triumphs and relationships forged across racial and class divides; of poor whites finding their way into middle class; of black women gaining momentum, by the mid 1960s, on their white counterparts and using their high school education to pursue professional degrees in higher education.

The darker story of the Border-Crossing Generation of the 1970s and early 1980s told of marginalized black and work-release youth who were ignored by school authorities; of the continued humiliation of the very poor; of coaches' abusive ways against young men; the egregious affairs of some male teachers with their female students; of the privileged status accorded by school authorities to jocks and the smart students at the expense of most other students across racial divides; and the humiliating labeling of youth by teachers as "losers." The brighter story of the Border-Crossing Generation told of friendships sealed between white and black young women and men on sports fields, in music bands, and classrooms; of interracial dances and couples; of white youth's vigilance against racism and commitments to egalitarian relationships across class and race boundaries; and of the social mobility afforded black youth through life-lasting connections established with wealthier white friends.

The darker side of the Redivided Generation of the late 1980s and 1990s told of student rivalries and prejudices along racial, class, ethnic, national, and religious divides; of fights erupting in the cafeteria and parking lots between Russian, black, and white racist youth, female and male; of stressful and fierce competition in the upper academic tracks for admission to colleges; of ready accusations of discrimination against teachers by students, and against students by students; and of broader community segregations that pitted old-time black and white families against newcomers to Miller Town. The brighter story of the Redivided Generation told of substantially increased social mobility for many African American youth whose economic status gained parity with the wealthiest whites in the community; of continued interracial friendships forged on sports fields; and of interracial and cross-national friendships forged among the wealthier preps.

* * *

Linking the Particular to the More General: Point and Counterpoint

That this history has unearthed divisive tendencies of the high school is not new knowledge. The comprehensive high school's systematic sorting of students has been well documented in educational histories and ethnographies, often linked to broader capitalist and market economic forces and examined in terms of institutionalized racism.[7] That this history has further revealed the role played by students in importing cultural influences from outside of school is also not knew knowledge, but is a phenomenon explored by anthropologists and educational ethnographers, although barely by historians of education.[8]

While this analysis confirms findings regarding the nefarious effects of tracking and students' economic and cultural realities imported into school spaces, it further expands our historical knowledge of student life and intergroup relations in high school beyond the historical implications already discussed in this conclusion, and beyond this work's extension of Gerald Grant's story into the end of the twentieth century.

First, this history reveals the impact of institutional norms on student relationships across dimensions of race, class, gender, and ethnicity, rather than on student achievement across those dimensions. As one researcher noted, "[W]hile student achievement data are critical, schools do more than teach academic subject matter: They have a profound potential for shaping individuals and their social networks."[9] In turn, social networks might play an equal if not more important role than achievement in post-school life opportunities. Jencks et al. found that "academic achievement does not have the kind of overwhelming impact on later occupational success that might justify making it the exclusive focus of most research on the outcome of desegregated schooling."[10]

Secondly, because this history brings to light the interplay across a substantial half-century, between school, demographics, social context, and citizens' lived experiences, it reveals across time institutional areas of intensified segregation and of facilitated integration": places of greatest vulnerabilities to divisive social forces and of greatest capacity for fostering authentic integration independent of broader social forces. Across the past half-century, as earlier discussed, Miller High's areas of intensified segregation proved to be the lower academic tracks and the large gathering space of the cafeteria, where conflicts heated up. Miller High's areas of integration proved to be the upper academic tracks and sports fields, from which students emerged into adulthood with habits of association that predisposed them to bridge divides—spots removed from heated conflicts. Miller High's tracking system, whether understood critically as a capitalist social-control mechanism and the tool of institutionalized racism, or benignly as the high school's democratic effort to "serve the needs and interests of all in a diversified curriculum,"[11] created school spaces that more readily divided or united students.

Historians have suggested that in the early twentieth century, high schools strove "to assimilate and amalgamate [the myriad diverse youth]...to Anglo-Saxon conception[s] of righteousness, law and order, and popular government"; that they "socialized youth to modern bureaucracy and connected their most intimate identifications to the power of the state"; that, in sum, they "allowed students to imagine that they were part of a national community."[12] The case of Miller High shows that by the end of the twentieth century, the comprehensive high school lost this purchase to amalgamate the students in its charge (if it ever truly had it—a subject for another study). This comes to view most vividly in the rejection of the "American way" by segments of the Russian and African American students, and in the racist attitudes of segments of native white students as well as segments of the Russian immigrant youth.

Under a half-century of changing demographics and a progressively more complex cultural environment, Miller High continued essentially to operate as its early twentieth-century counterpart. The case of Miller High suggests that while the comprehensive high school of the end of the twentieth century continued to track and sort citizens—business as usual—it became, within the new world order following desegregation, the technology that tightened divisions along religious, ethnic, and racial identities in ways that intensified animosities across divides, rather than fostered respect across those divides. Moreover, while the comprehensive high school of the early twentieth century had also been "the indispensable technology that helped reorder the relationships between self and society...[and] loosen the grip of local, religious, ethnic, and racial communities,"[13] the case of Miller High suggests that by the end of the past century, it became the very technology that splintered U.S. citizens and residents into diverse local loyalties.

By the end of the twentieth century, a very diverse student body had been sensitized to issues of discrimination, the case in point being Miller High's recently immigrated Russian youth who openly accused "American" teachers of discriminating against them, something that immigrant youth of the nineteenth century would have been less likely to do. Another case in point is Miller High's recently migrated city youth, who had been much more attuned to racism than the local black population. While at the turn of the twentieth century the comprehensive high school might have been a response to the perceived threat posed by diverse youth ready to compete for jobs with adults, and while it became the technology that kept youth off the streets and served to assimilate them into the order of the day, as Joseph Kett and other historians have suggested, by that time it had lost such authority. The early twentieth century "threat" of diversity had been dislocated from the street into the school by the end of the twentieth century. Diversity's "threat" to marketplace social order had developed over time into a threat to "school business as usual." Miller High's Anglo middle-class normative practices and its institutionalized regimentation and herding of student bodies turned "diversity" into a problem rather than a solution.

At the meeting place between a racially, ethnically, and religiously militant youth and the rigid institutional normative habits of tracking and herding, for school authorities, social order seemed threatened within school walls. Unimaginatively, school authorities responded by becoming more punitive. However, the reader will remember how by the 1990s rule-abiding students in lower tracks preferred strong autocratic disciplinary measures that made them "feel safer," protected against law-breaking students, who often happened to be Russian, black, and "rednecks." The reader will further recall how by the 1990s divisions among students created parallel worlds, a citizenry divided not only by race, ethnicity, and class, but also as rule breakers and rule abiders. One can reasonably claim that the dark side of this history presaged the rise of a citizenry accustomed to a policing government. The case of Miller High shows a punitive system sustained by the fears of rule-abiding Miller High students, further fueled by aggressive and mistrustful disenfranchised students and a rigidly divisive tracking system.

Miller High had increasingly opted over time to police students—to keep order under business as usual. In considering this darker history, one might further venture to presage the future weakening of democracy, witnessed in the increased policing of citizens at the dawn of the twenty-first century (e.g., Patriot Act, wiretapping), a policing that has been examined only sporadically and with delayed reaction. One might also venture to presage, as we globalize the comprehensive high school, the making of a multiracial, multiethnic, economically advantaged minority (the multicultural preps), and a multiracial, multiethnic, economically disadvantaged majority, as we "stuff" more students into lower tracks. These sinister, science fiction—like notions, while too simplistically and too harshly forwarded in these concluding remarks, are not unimaginable if the comprehensive high school is to continue business as usual as it has since its inception.

There is, however, a counterpoint to the dark story. When looking at the brighter side of this history, captured in the Border-Crossing Generation, one can imagine a civil rights renaissance and a student body that once again reaches across divides. One can imagine a youth buttressed by a secondary public school system architecturally designed to create smaller spaces for students to interact across differences. One can imagine a vocational or apprentice system incorporated into the workplace and attended by students following graduation from high school and not during high school. One can thus imagine dismantling the "comprehensive" in the comprehensive public high school and opting instead for a liberal arts curriculum for a twenty-first century global reality. A shared liberal arts curriculum would do away with the low status lower academic tracks within which, as this history has shown, racial and ethnic divides are intensified. Moreover, a shared liberal arts curriculum might also help lessen class divides since lower tracks are mostly holding places for students of lower socio-economic background. Finally, one can imagine a constant presence by a diverse body of school authorities among youth, one that exemplifies relationships across divides in everyday interactions, rather than a policing, distant administration. One can then imagine a fresh generation of youth that gives birth anew to a U.S. democracy that best exemplifies a "world lived in common with others."

Addendum Methodology: The Transparent Historian

The purpose of this section is to describe and discuss the challenges for data collection and interpretation of what is known among historians engaged in collecting oral histories as "oral history doing."[1] I begin by addressing the use of the guiding theoretical framework intersectionality in collecting the oral histories for this analysis. Historians de rigueur abhor theory, and for valid reasons, but I argue, in what follows, for the relevance of this particular theoretical lens for this particular oral historical research. I then address the more technical aspects of oral history doing: identification and location of graduates, interview protocol, transcribing, and coding. I end by describing my archival research and the particular analysis of yearbooks for this history.

Theory and Oral Histories

I would like to begin by emphatically stating that oral historians not only reproblematized memory for historians by focusing attention on understanding the subjectivity of memory as a manifestation of historical consciousness,[2] but that they also brought to our attention how memories are gendered, racialized, and class based. They brought to our attention the importance of examining why different individuals and groups experience the same event in very different ways.[3] Today, oral sources—particularly since the 1990s, when historians began investigating the construction of identities[4]—are compared less pejoratively to documentary sources by academic disciplinarians. Additionally, oral historians continue to break down "boundaries between the educational institution and the world, between the [history] profession and ordinary people."[5] However, they also continue to be faced with the challenge of articulating "the connection between individual and social historical consciousness."[6]

This theoretical challenge—transforming individual memory into social memory—becomes particularly relevant to those of us interested in the production and analysis of oral testimonies of school life across generations in the United States, since desegregation, as these reflect on peer relationships across socially constructed categories of difference (race, class, gender, religion, nationality, etc.). When, how, and under what circumstances do individual memories of school relationships credibly tell a social rather than psychological story? How does one determine the number of testimonies necessary to legitimate a social rather than personal history?

To this theoretical challenge is added the *problématique* of the relationship between interviewer and interviewee, professional historian and narrator. When oral historians stir up memories of school experiences, they stir up worlds of identities in the making. They also stir up present-day integrated meanings of school and the educational system, meanings that simultaneously interlace, buttress, and challenge memories. Who speaks: the adult reconstructing the past to justify the present, to justify the past, and to whom? To herself or to the historian? How to deal with the power-laden nature of co-creating with the narrator a primary source—that is, the very data one uses to build one's analysis?

Fundamental to the theoretical and methodological challenges posed by researcher-participant co-constructed interviews is the issue of language itself. As a former foreign language teacher and avid student of sociolinguistics, I construct voices as social rather than primarily intrapsychic phenomena. In this view of language, meanings are constructed by individuals as these individuals relate to one another.

"The word in language is half someone else's... It is populated—overpopulated—with the intentions of others... to understand voice, researchers must accept that what they hear is a function of who they are as individuals within the social community."[7] Thus the oral historian contends in her métier with Bakhtin's heteroglossia; that is, she navigates the many voices situated within dialects, generations, class, race, and national consciousness; she contends with the diversity of voices as they speak against or to each other, and as they speak differently or similarly to her own. The integrity of her work rests not only on examining critically the linguistic register of collected oral testimonies as well as archival documents, but also on critically examining the sociocultural intentions revealed by her own linguistic markers. The essential nature of oral historical inquiry naturally and relentlessly brings the oral historian to theory.

Broadly, theoretical frameworks provide needed discursive templates to talk about the challenges presented by oral historical work that seeks to examine how diverse students adapt to multiple cultural contexts and affiliations within single institutions over time. One of the top challenges is the obvious intrusion of the historian in the production of oral histories. The consequences of that intrusion often create metadata that cannot be easily tucked into the narrative, but that should not be relegated to footnotes. These are data that interrupt the narrative. I call metadata those bits of information that do not speak directly to the past that I seek to recover, but to narrators' present reactions to and reflections about the oral historical research process. These reactions often shed important light on narrators' developed historical consciousness about the story being researched. Theory helps organize human phenomena in their relation to one another. In the case of my work, intersectionality theory becomes a useful lens for the complex stories that I have to tell.

Black feminist theorist Patricia Collins's intersectionality theory offers a framework for positioning both historian and narrator within the "doing" of oral history. It also offers a framework for capturing alumni's shifting constructions of cross-group relationships within and across time periods.[8] The intersectional analytic approach examines the ways in which social markers of difference (race, gender, ethnicity, sexuality, generation, class, religion, and nationality) intersect to shape situated experiences, and places the historian within her own social position. Intersectionality challenges homogeneous constructions of social groups and brings to light the complexity and fluidity of group identities. The theory grew out of black feminists' articulation of the limitations of "gender as a single analytical category";[9] and by extension, the limitations of any single social category to capture the complexity of cross-group relations.

My application of the theory addresses both the intercategorical and intracategorical dimensions of intersectionality.[10] The intercategorical dimension focuses attention on the relationships between social groups and identifies the power differentials at play in those relationships. The intracategorical dimension historicizes social categories and brings to light their shifting definitions by identifying people who "cross the boundaries of constructed categories."[11]

Intersectionality applied to the oral historical enterprise per force situates the oral historian not only socially, vis-à-vis individual narrators, but also historically, vis-à-vis changing understandings over time of the researcher's status and role. I begin then by situating myself to you, the reader, that you may consider the many implications of my arguments from your own situated perspective.

It is from the situated perspective of a naturalized U.S. citizen; of a white woman of hybrid cultural and diverse linguistic backgrounds whose race and sexual orientation accord her important social and political privileges in a society that favors whiteness and heterosexuality, while her gender and role as mother restrict her economic and professional opportunities;

and it is further from the perspective of a multidisciplinary scholar who was schooled in French Canada and Mexico prior to graduate work in U.S. universities that I engaged in conversations with diverse alumni who attended Miller High between 1950 and 2000.

In the history of Miller High, subjectivities created and recreated one another. Graduates' testimonies did not merge and rise into collective memories for each generation, but rather into shared interpretive frameworks regarding the nature of peer relationships across markers of difference. Thus from a mosaic of differently positioned individual experiences emerged the relational historical consciousness of Miller High generations. While demographic composition, institutional norms, and broader political and economic environments shaped the perimeters of possible student encounters across social categories for any given period; and while the voices of alumni reflected the general social grammar of their times, the history of cross-group high school relationships is grounded in graduates' testimonies—in individual, subjective experiences. Remembered testimonies were given primacy over documentary data; that is, archival data was used to contextualize testimonies, not the other way around.[12]

Furthermore, while I conducted archival research before, during, and following interviews to situate the school and its community, and to see if "facts" stated in testimonies had any grounding in broader social realities or official data, the archival research was not conducted to challenge or discount the memories shared, but to situate them more clearly when possible. The "truth status of individual stories became less important than the value they [were] trying to support in the telling"[13]—because what was sought was the lasting impact of school on the development of school relationships, and the impact of school relationships on perceptions of "others" as experienced by male and female graduates of diverse racial, ethnic, and class backgrounds.

For relational histories of education—such as this one, interpreted from Miller High alumni memories—no ultimate number of testimonies can be set to legitimate a social historical consciousness (over personal history). No number can be set because no exhaustive story of all possible configurations of student relationships can ever be told. It is because the "exhaustive" will forever be out of reach in relational histories of education that histories of whatever possible configurations must be told. They must be told because they begin our foray into the evolution and/or dissolution of intercultural, multicultural, multiracial, multiethnic, and religiously, economically, and nationally diverse educational spaces.[14]

Intersectionality theory offers a way to connect graduates' subjective interpretations of school relationships to their situated positions within both school structures and broader social contexts, and in reference to others' subjective interpretations. It is the very juxtaposition of differently situated perspectives within and across time periods that builds a relational historical consciousness. Thus the intersectional lens helps the oral historian identify and articulate how socially constructed markers of difference intersect within individual historical protagonists and how they shape nuanced group experiences that do not monolithically assign equal experiences to all blacks, all whites, all women, all men, and so on, within institutional bounds. It is precisely because this lens helps the oral historian identify human experience within intersecting categories of socially constructed identities that it also helps the oral historian to identify more clearly the fluidity across time of the very social construction of identity categories. Thus, rather than contribute to essentializing experience within social categories or to perpetuating categories of identity that are arbitrary, the intersectional theoretical lens helps more starkly bring to light how historically specific these constructs are. Intersectionality theory helped this oral historian frame and organize what continues to be a challenging history to grasp.

I repeat what I have already addressed in the Introduction: Informed by the theory of intersectionality and committed to an oral historical perspective that foregrounds the subjective experience of history—not to rewrite facts, but to gain a deeper understanding of how and why the same facts are reconstructed differently across historical protagonists—I forged ahead to find and tell the story of Miller High students' cross-group relationships from those perspectives recovered.

Identification and Location of Alumni

I identified graduates by searching senior pictures in Miller High yearbooks. While overall and broadly I could identify graduates in yearbooks by gender and race, and even to some extent by ethnicity through last names, I could not as easily identify them by economic status; serendipitously, I fell upon graduates of different economic backgrounds (more about class identification follows). When graduates identified through yearbooks could not be located in the alumni directory, I forged ahead until matches were made. It was particularly difficult to locate African American graduates of the 1950s through the 1980s, and in particular men, who rarely included their names in the directory, a telling story of its own.

Doors opened up for me into the African American community when I came across a history, in the town library, written by Louis Diggs, an

African American oral historian who had collected the testimonies of members of some of the oldest African American communities in Maryland.[15] I contacted the author, who suggested I speak with Annie Milligan, an octogenarian who had lived her entire life in the African American community that Miller High served. Mrs. Milligan took me into her world and connected me with African American Miller High graduates. It was her trust in me and her standing in her community that secured my interviews with African American alumni who attended between 1956 and the early 1980s.

Through Mrs. Milligan's intercession I gained invaluable insight into Miller High's black community of the 1950s through the 1980s, but I could not do the same for the younger generation of the 1990s. As discussed in the body of this work—while those alumni interviewed freely agreed to give their stories to me, the implications of my whiteness within a U.S. reality, in conjunction with my PhD candidacy in education at the time of the research, and the power differential inherent in the production of research interviews, created in the eyes of two potential African American narrators, 1990s graduates, a person to whom they would not want to grant an interview. The agency of these two alumni was recorded within the narrative. They refused to give their insights and interpretations to someone whom they understood to represent an educational establishment that has failed black students. Their political statement reconstructed the power differential on their behalf. Their refusal to tell of their experiences is a historic act that was recorded as clearly as remembered testimonies. It could have been omitted or—perhaps worse in this case—footnoted. Their refusal is part of the metadata that deepen and complete the history of Miller High student relationships, but inconveniently interrupt the narrative.

By adopting the intersectional theoretical framework and abiding by its logic, one risks exposing one's vulnerability as a researcher. Mine was exposed, and in the process, a historical defiance was recorded, one that must be understood positively—the agency of real people in the face of institutionalized and institutionally sanctioned educational research that lacks immediate transformative power.

My situated persona also elicited a response from one white alumnus, which brought to view a parallel story of "old-timers" and "newcomers" in the Miller High community. Tim Whittle, a white alumnus of lower socioeconomic status when attending Miller High between 1974 and 1981, expressed an opinion about how I might be perceived by the town's "old-timers" at the time of the interview:

> In this huge town, of the people who are here now—as my grandmother used to call them, the 'move-ins'—she wouldn't even talk to you! You

"move in" with an accent! [Tim imitates his grandmother's indignation, and laughs, razzing.]

In this case, it was not my whiteness, but my foreign accent that created an opportunity for the alumnus to make a statement about a social hierarchy of belonging. Tim's comment alerted me to the possibility of a parallel story—a story that emerged from Tim's reaction to my linguistic register. Indeed, it became the story of cross-generational strife in the community between long-established black and white Miller Town families and newly arrived immigrants and black youth from Baltimore throughout the 1990s. But in the live human encounter of the oral historical interview, a metadata set also emerged to complicate the story: the fact that the researcher's accent made her suspect. Intersectionality compels one to reveal the very encounters with narrators that tell of the import of the situated position of each in the making of historical evidence and interpretation. Such transparency serves my work well.

My goal was to identify an average of ten graduates per decade, and to include as much as possible equal proportions of male and female graduates of different racial and economic backgrounds who attended during the first and second half of each decade. I decided on ten graduates per decade given my resources at the time of the research. Since my intent was primarily to locate different perspectives and to capture meanings ascribed to relationships in school rather than investigate the veracity of a past occurrence, difference, not volume, mattered. Secondly, grounding my study deeply within a single institution made collecting a smaller number of oral histories much less problematic than if I had spread my study across numerous sites. Finally, I reiterate that I was not interested in producing a "collective" memory of Miller High students over the second half of the twentieth century. Undeniably, such an endeavor would have required "volume" and would have carried headaches of its own, since by the end of the century, fragmented perspectives (as this work has shown) would have challenged the notion of a "collective" memory. Seeking to construct a "collective memory" works best when the group whose memory is sought is homogeneous.

Although I had originally scheduled to interview forty-five graduates, I ultimately interviewed thirty-seven. Eight canceled and I ran out of resources to pursue further interviews. Beyond the Russian population of the late 1980s and 1990s, immigrant students also included Asian Indians and Latinos. They, however, represented a significantly smaller community than that of Russian immigrants at the time, and again, while it would have proven invaluable for this study to include their perspectives (the absence of which I sorely lament!), I did not have the resources to invest in locating and interviewing these populations of immigrants.

I began the data collection process by sending letters to identified graduates. I included an explanation of my project along with a stamped return envelope for their reply. Five out of the fifty people I had written to in my first batch of letters answered my mail; two declined to be interviewed, and those who answered were all graduates of the 1950s. Of the second batch, none responded. The three graduates of the 1950s who did accept proved to be invaluable contacts, as did Mrs. Milligan, through whom I contacted several African American alumni who then led me to other alumni, black and white. Thus I located graduates through purposeful sampling as well as snowballing.

Graduates of the 1950s were much easier to locate and much more eager to tell their stories, so much so that as a single researcher, I could not afford to interview them all. While I could locate later graduates, they were not as eager to be interviewed, being busy with work and raising children. This explains the disproportionate number of graduates, and of female graduates in particular, interviewed from the period before desegregation in the 1950s when compared to the other decades. However, their perspectives served to establish a baseline, a construction of student relationships among all white youth immediately preceding the desegregation of Miller High in 1956. Their perspectives allowed me to capture the already institutionalized segregating forces at Miller High.

I proceeded with interviews as soon as I could schedule them, even as I worked to locate others. In the early stages of data collection, interviewing and locating graduates to be interviewed proceeded in tandem. While I had planned to begin interviewing graduates of the 1950s and of the latter part of 1970s first, since they were located at the beginning and middle of the time period—to feel out the decades—my plan soon dissipated. I had to adapt to alumni's schedules and availability. Nevertheless, I was able to spread out my beginning interviews across the decades and to pick up somewhat on changes in mood across time periods.

Interviewing Protocol

After introducing myself and explaining the project, I took time to make small talk while I prepared my recording equipment and to set narrators at ease about the taping process. I emphasized that we could stop the recording at any time (for a break, if something needed to be said "off the record," etc.), and I explained that they could choose to donate their recording to the town library now or posthumously, but that I would keep their identities confidential for the purposes of my work. Because I had previously conducted pilot interviews for which graduates requested anonymity, I learned that it would be best that all participants sign a letter of consent that underscored the confidentiality of their testimonies for the purposes of my study. Too many graduates are still active members of the Miller Town community and did not want to compromise their current relationships with comments made about the past that would become public. Some have agreed to donate their recordings to the historic room of Miller Town's library posthumously. To protect agreements of confidentiality, all proper names, including the name of the town and school, have been changed to pseudonyms. After long deliberations, I was satisfied that situating Miller Town within Baltimore County and Maryland would not compromise graduates' identities, since there are several high schools comparable to Miller High in the county, and since the influx of Russian and other immigrants affected several high schools, not only Miller High.

Thus, beyond an explanation of the purpose and procedures of the study, the consent form also included the following:

> All information collected in this study is confidential to the extent permitted by law. I understand that the data I provide will be grouped with data others provide for reporting and presentation and that my name will not be used. I will be given the option to donate my tape recording to the [name of town] library archives. If I wish to do so, I will sign a secondary release form for this purpose.

I began all interviews with open-ended questions that asked graduates to describe their relationships with peers, teachers, and authority figures. This proved a very productive way into the conversation for all my interviews because it provided the space for graduates to jump into their memories as these came to them, without my imposing a prior order. As graduates' comfort levels increased, I followed with more probing questions about details of their experiences and the meanings they ascribed to their high school education and diploma. Only twice did such a broad, open-ended question spur graduates to meander on tangents into their personal lives, tangents that had nothing to do with their high school experiences. Nevertheless, allowing them to meander helped establish a rapport that yielded fecund interviews. While some researchers send a list of interview questions in advance to help the informants prepare for the interview, I chose not to send questions in advance. I wanted the conversation to emerge spontaneously, and I wanted to avoid formally written questions that might restrict either alumni or me.

Thirty-two interviews were held in graduates' homes, mostly during weekends or in the evening after work; one interview was conducted in the lunchroom at an alumna's workplace; three were conducted in a quiet area of the archives in the town library; and one was held at a bookstore. I always gave graduates the choice of our meeting place; I also took precautions, especially when interviews were conducted in homes and at night, by arranging to be picked up by a family member. Overwhelmingly, graduates and I quickly established good rapport.

Tape-recorded sessions ranged from one to two hours. On several occasions I followed up formal interviews with informal chats by phone or e-mail. To the thirty-seven oral histories conducted with alumni, I added one group interview with three Miller High teachers. One teacher, a female coach, had been working at Miller High for thirty years. Between them, the Spanish and English teachers had fifty years of tenure. The opening question I posed was to describe their experiences of Miller High students over time. This interview lasted fifty minutes and was conducted between classes in a quiet room of the Miller High library. Finally, I also interviewed, but did not record, the principal. The meeting lasted half an hour. See Tables A.1 and A.2 at the end of this section for a list of interviewees by time periods.

About Class Identification

The social constructs "middle class," "poor," and "working class" are tricky categories within a U.S. reality, and even more tricky when one crosses racial borders. What constitutes a middle-class range for a white citizen might not constitute the same range for a black citizen across historical periods and geographic contexts. It is important to underscore that I did not a priori choose a sociological definition of class as I engaged in my interviews. Class identifications emerged organically as narrators situated themselves and/or others they remembered in contrast to their positions on a continuum of "poor" to "middle class." I reported narrators' economic status as they described it. Nine narrators did not identify what their economic status was when they attended high school. In those cases I broadly deduced their class backgrounds by identifying their childhood homes in the town, the jobs they entered following

Table A.1 Pseudonyms arranged by generations, graduation dates, and interview dates

Divided Generation (1950–1969)	Divided Generation (1950–1969)
Before Desegregation (1950–1956)	**After Desegregation (1956–1969)**
Judy Law (1950–1954) / 02–25–2003	Doris Right (1956–1958) / 9–4–2003
Dorothy Kaufman (1950–1954) / 02–04–2003	Annie Cole (1956–1958) / 03–11–2003
Alice Web (1950–1954) / 10–02–2003	Norman Good (1956–1959) / 03–18–2003
Sandy Eycke (1950–1954) / 03–18–2003	Sherry Parson (1958–1962) / 03–07–2003
Robert Heart (1952–1956) / 6–24–2003	Nat Right (1956–1963) / 07–19–2004
Lou Anne Kensington (1953–1957) / 06–12–2003	Dotty Morris (1961–1965) / 09–24–2003
Nora Jones (1955–1959) / 06–12–2003	Burt Sadden (1963–1967) / 07–26–2004
	Betty Land (1964–1968) / 02–08–2003
	Linda Moss (1965–1969) / 09–24–2003
	Bud Land (1965–1969) / 08–16–2003
Border-Crossing Generation (1970–1985)	**Redivided Generation (1986–2000)**
Joanne Pet (1970–1974) / 10–16–2003	Harry Rice (1983–1987) / 02–25–2003
Sam Garnes (1970–1974) / 09–16–2003	Roberta Jones (1987–1991) / 09–02–2003
Pat Baley (1971–1975) / 10–20–2003	Cherry Gate (1988–1992) / 10–23–2003
David Randle (1972–1976) / 11–07–2003	Heather Korn (1989–1993) / 02–25–2003
Josh White (1972–1976) / 10–01–2003	Sue Cohen (1991–1995) / 09–2003
Michael Hallner (1972–1976) / 9–2003	Cecilia Hood (1992–1996) / 10–7–2003
Jeremy Garnes (1974–1978) / 09–09–2003	Betty Ames (1995–1999) / 08–19–2003
Timmy Whittle (1977–1981) / 06–28–2004	Bill Jackman (1995–1999) / 10–18–2003
Teresa Randle (1977–1981) / 06–09–2004	Vera Debin (1995–1999) / 01–15–2004
Sophie Baker (1981–1985) / 06–17–2004	Ivan Strasky (1998–2002) / 11–06–2004

Note: Interview with teachers was conducted on 11/07/2003: Michelle Shaw (English)—began teaching at Miller High in the 1970s; Julia Mills (coach)—began teaching at Miller High in the 1960s; Sam Buck (Spanish)—began teaching in the 1970s.

graduation, or colleges they attended. For example, if a narrator testified to having worked after school "to survive" and reported taking a job at the local factory after graduation, I deduced that the narrator broadly belonged to the working class at the time he attended high school. See

Table A.2 Distribution by race, gender, and class (at the time of school attendance) of graduates across generations. "NI" indicates that graduates did not identify themselves by class. Class identifications are reported in language used by graduates

Divided Generation (1950–1969)

Before Desegragation (1950–1969)

— 1 white woman / poor
— 6 white women / middle class: 3 of them were college-bound, suggesting greater financial wealth than the other 2 who stated not having been able to afford college.
— 1 white man / poor

All white graduates
7 women
1 man

Divided Generation (1950–1969)

After Desegregation (1956–1969)

— 3 African American women / middle class: 1 lived in an upper-middle-class house in a richer neighborhood, and 1 worked after school to help family financially.
— 1 African American man / rich
— 1 African American man / middle class
— 1 African American man / NI: he had to work after school "to survive" suggesting perhaps lower middle class / working class
— 2 white women / middle class: both were college-bound and part of the "doers" identified by later generations as the "preps," suggesting the higher end of middle class
— 1 white woman / middle class: but, she could not afford to go to college
— 1 white man / middle class: he was going to college and his father owned a lucrative business suggesting higher end of middle class

6 African American graduates
4 white graduates
6 women
4 men

Border-Crossing Generation (1970–1985)

—2 African American women / middle class
—2 African American men / NI: their testimonies about where they lived when attending school suggest the broad category of middle class
—5 white men / NI: testimonies of one of them suggested that his background was working class and had worked in a factory after high school; I know where the other graduates lived when attending high school suggesting a middle-class background. The father of one of them was a small business owner.
—1 white woman / middle class

4 African American graduates
6 white graduates
3 women
7 men

Redivided Generation (1986–2000)

—2 white men / middle class
—4 white women / middle class: 1 was college-bound while the other 3 were very sensitive about not having the consumer power that the "preps" held; one of them in particular identified as a "farm girl" and her husband had been a work-release student at Miller High who later became a mechanic.
—2 women / middle class: preps, suggesting upper end of middle class
—1 white Russian immigrant man / NI: the house where he grew up and in which I interviewed him was upper-middle class (neighborhood where houses sold for 500,000.00 dollars in 2003)
—1 white Russian immigrant woman / middle class

0 African American graduates
10 white graduates, of which 2 Russian Immigrants
7 women
3 men

Note: One African American and one white woman, graduates of 1972 and 1985, spontaneously shared during their interviews that everyone at Miller High was middle class.

Table 2 in this section for a summary of narrators' class identifications by gender and race.

Transcribing and Coding

As much as possible, I transcribed interviews as soon as possible. Each transcription yielded on average twelve single-spaced pages. When vocal inflections or laughter qualified a remark in a striking way, I made note of it in the transcript. I spent an average of seven hours of transcribing per one hour of recording. I proceeded with analysis alongside data collection as I transcribed. The very act of transcribing was for me, in many ways, an important part of analysis. It was during the transcribing process that I jotted notes about comments I had not probed to my satisfaction, to revisit with the alumnus. Transcribing became more than just a preliminary stage to coding and analysis; it was the first stage of analysis as graduates' voices began echoing or contradicting each other. Transcribing kept the voices loud and alive in my mind.

I coded manually, without the use of a software program;[16] and I coded chronologically, beginning with the transcripts of those alumni who graduated in the 1950s, then the 1960s, and so on. Within each decade, I first created profiles for each alumnus, after which I looked for patterns across graduates' experiences. I did not a priori assume that being black or white or female or male, or other, would per force create similar experiences within those categories. It was important that I allow identifications to emerge from students' remembered lives, within their time and place, as they may. In the process, I developed my own idiosyncratic coding style.

For each transcript I created accompanying coding sheets on which I identified the alumnus, and where I organized themes that I had identified for the particular transcript. All of my coding sheets were written by hand, in pencil, allowing me to easily rearrange themes; these were attached to transcripts. Coding sheets were first completed for each individual transcription and then compared across transcripts within decades. Transcripts that shared most themes and shared language/linguistic register were grouped. From there, my analysis proceeded in narrative form. I compared experiences between groups within decades and then across decades. At this stage the narratives were two or three pages long per group and mostly written in syncopated prose. It is during this last stage of comparison (across decades) that generations emerged as groups consolidated.

As discussed in the Introduction, the echoed memories of student relationships as reported by those graduates interviewed across race, class, gender, and other divides, as well as memories collected through informal interviews over a two-year period (2003–2005), suggested that enough of those Miller High graduates interviewed perceived cross-group relationships similarly within certain time periods, as well as differently across time periods, to constitute generations; and that enough of them articulated in similar language those similarities out of which emerged time periods, which were further corroborated by broader demographic changes in Miller Town. Bound by shared memories of institutional practices, shared demographic configurations, and shared articulations (in use of vocabulary and cultural references) of cross-group relationships, the oral historical testimonies sketched the perimeters of generations.

It is also important to underscore that when certain voices were missing—for example, African American voices of the 1990s—their presence came into view obliquely through the testimonies of those interviewed, and were further brought out of the shadows, as much as possible, through yearbook and other documentary data and conversations with Miller High residents, black and white, over a two-year period (2003–2005). Sadly, Latino, Asian-Indian, and other voices of the 1990s are missing completely—the focus of further research.

Memory-Elicited Data

Within the world of memory, experience has been interpreted, perceptions have been sifted and settled, and meanings have been attributed. Within memory lingers the lasting impression.[17] Evidential issues surrounding the question of memory, while sensitive, held less weight in this analysis than they might have had I attempted to investigate, for example, the reputation of a particular Miller High teacher or administrator through students' remembered oral testimonies; or had I sought to construct an overarching collective memory of a homogeneous group of people with shared interests in representing their reality unilaterally.

Still, one might critique use of recollections as a means to capture students' interpretations since the interpretation of one's experience while experiencing it may have changed as one gained maturity. What circumvents this problem, in this study, is the live coproduction of the interview, which allows for a distinction in the present context, during the interview making, between the interpreted experience at the time of the memory making, and the reinterpreted experience a posteriori. For example, one alumnus recalled thinking that homework was useless, but today thinks that homework in general might have some grounding value. Of interest to me, within the context of this research, and regardless of the alumnus's present-day reinterpreted evaluation of homework, is that when he was a student, he thought homework was useless. Sensitive probing usually reveals the weight of present interpretations of past experience, and informants are usually interested, and in some cases even relieved, to explore those differences.

In this sense, the making of an interview allows for analytic distinctions that cannot be as readily made when mining such sources as diaries, journals, or biographies, where a posteriori reinterpretations are more easily slipped into the narratives by those who lived the history and are not as readily evident to the researcher. Moreover, it is in the comparing of experiences across situated perspectives that I gauged to what extent one graduate's understanding, at the time of his/her school attendance, might have been an idiosyncratic understanding and to what extent it was shared and with whom.

Archival Research

Archival research was conducted in the Baltimore County Board of Education archives in Towson, Maryland, where general county school policies were examined; in the Maryland rooms at Enoch Pratt Baltimore City Public Library and the University of Maryland College Park, where Maryland State Board of Education communications and census data were examined; in the library of Miller High School, where the school newspaper and archived alumni achievements and photographs were examined; and the "historic room" of Miller Town community library, where Miller High yearbooks and local community newspapers were examined. In the Miller Town library I also referenced histories of peoples of Baltimore County.

Locating records of student populations proved impossible. Miller High does not keep information older than five years (when I visited, the school had undergone a recent "major cleanup," as the administrative assistant explained), and the information available at the board of education tracked population shifts only by geographic sections, not by individual schools. Therefore, I tabulated student populations by counting senior pictures in yearbooks, which I compared against county and Miller Town census data. Moreover, various additions to the 1930s building, documented in the *Alumni's Directory*, further confirmed testimonies of changes in population at Miller High. Finally, I acquired a good sense of population changes through editorial comments expressed in the *Community Times* over time, as well as through my own encounters with the people of Miller Town between 2003 and 2005.

Of particular interest to this study is the use I made of yearbooks, inspired by the work of youth historian Paula Fass, who, in her essay "Creating New Identities: Youth and Ethnicity in New York City High Schools in the 1930s and 1940s,"[18] opened up new ways to consider yearbook data. Fass analyzed the distribution of students by ethnicity in various extracurricular activities. Following her lead, I developed a systematic record keeping of the distribution of students by gender, race, and ethnicity in the various extracurricular activities from yearbooks for every other year beginning with 1954.

For example, when averaging student participation by gender and race for the student "yearbook club," "newspaper club," and "honor society" (for yearbooks dated 1954, 1956, 1958, and 1960, 1962, 1964, 1966, 1968), the following proportions emerged: on average, thirteen white female students compared to eight white male students served on the yearbook staff, and no black students served across the two decades; on average, twelve white female students compared to five white male students served on the school newspaper staff, and no black youth (only one anomaly appears in 1964, when there were only ten white female students compared to fourteen white male students); on average, 126 young female students compared to seventy-eight white male students, and no black students, were in the honor society across the decades. Future Teachers of America and Future Nurses of America were consistently dominated by white women (only one white male appears in the late 1960s in Future Teachers of America, and two African American females appear in Future Nurses of America in the middle and end of the 1960s). Across the two decades, only women are represented in the Future Business Leaders of America across races, the extracurricular activity that recruited young women from the commercial track preparing to become secretaries. Populated solely by males were the Future Farmers of America club and, in the late 1960s, the "chess club." Also, young black men were enrolled in basketball and/or track, and of those most were also attending academic tracks (this could be easily corroborated with senior graduating pictures that included students' academic and extracurricular credentials). Football would not show up until 1969. There are no sports pictures featuring female African American students between 1950 and 1969. They do appear in the 1970s and thereafter. Such analyses were conducted for all the decades. Also, see Table A.3 at the end of this section for an organized tabulation of Miller Town population growth derived from census data.

I would like to end this section by reiterating the relevance of intersectionality theory in oral historical research that explores generational

Table A.3 Population Growth in Miller Town

	Total	White	Total Population	
			Identified as Other than white	Identified as Other than white ~ %
1950	2,077	1,912	165	~ 8%
1970	14,037	13,724	313	~ 2%
1980	19,385	16,928	2457	~ 13%
2000	22,438	16,467 **Russian Population:** (1413) **6.3%**	5971	~ 28% Black or African American: 18.4% Hispanic or Latino of any Race: 4.4% Asian: 4% Other: 1.6%

Source: Census Data / Baltimore City Enoch Pratt Free Public Library, *Maryland Room*
U.S. Department of Commerce Bureau of the Census 1950, Vol.1, Chapter 20
U.S. Department of Commerce Bureau of the Census 1970, Vol. 1, Chapter 20
U.S. Department of Commerce Bureau of the Census 1980, Microfiche / Tape File 3A / Place: <Miller Town>

transformations in the relational experiences of youth attending desegregated schools in the latter part of the twentieth century. I suggest that theory lurks in any intellectual inquiry—acknowledged, its influence can be argued; and the historian lurks in any historical inquiry—acknowledged, intellectual authority can be redistributed.[19]

Notes

Preface

1. See James Anderson and Dara Byrne, eds., *The Unfinished Agenda of Brown v. Board of Education* (Hoboken, NJ: John Wiley & Sons, 2004), for discussions regarding the legacy of *Brown*.
2. For a discussion of school desegregation policies and competing arguments regarding *Brown v Board* versus the *Coleman* report, see Amy Stuart Wells et al., "How Society Failed School Desegregation Policy: Looking Past the Schools to Understand Them," *Review of Research in Education*, 28 (2005): 47–99.
3. Wayne Urban and Jennings Wagoner, *American Education* (New York: McGraw Hill Companies, Inc., 2000), 288–289. Urban and Wagoner suggest that *Brown v. Board* "may have been one of the few occasions in our history when an educational policy was the catalyst for substantial changes in social relations and policies outside of school." Certainly, a series of legislations followed during the 1960s to enforce the resisted implementation of *Brown v. Board*—legislations such as the *Civil Rights* Act of 1964 that would indeed affect social relations outside of school. However, as some historians have argued, the political and judicial changes that ensued during the civil rights movement following *Brown v. Board* cannot be easily or primarily attributed to *Brown v. Board of Education* and must take into account the pre-*Brown* struggles that laid the foundation for the civil rights movements. For example, see John Dittmer, *Local People: The Struggle for Civil Rights in Mississippi* (Champaign, IL: University of Illinois Press, 1995). See also Jack Dougherty, "From Anecdote to Analysis: Oral Interviews and New Scholarship in Educational History," *Journal of American History* 86 (September 1999): 712–723, for reticence by many black communities to embrace school desegregation. For resistance to desegregation in the north, see the classic work by Anthony Lukas, *Common Ground: A Turbulent Decade in the Lives of Three American Families* (New York: Vintage Books, 1985).
4. "The 1965 Amendments to the Immigration and Nationality Act: repealed national origins restrictions which restricted immigration visas according to

the ethnic composition of the 1920 U.S. population. Before 1920 the United States had a very high rate of immigration. The 1965 amendment essentially returned immigration policy to the pre-1920 policy; and the 1990 Immigration Act: permitted the entry of 150,000 more legal immigrants annually"; http://sorrel.humboldt.edu/~economic/econ104/immigrat/ (January 2006). See also Alejandro Portes and Rubén G. Rumbaut, *Immigrant America: A Portrait* (Berkeley and Los Angeles: University of California Press, 1996).
5. For an in-depth discussion of multiculturalism as it has shaped education in the United States, see James Banks and Cherry McGee Banks, eds., *Handbook of Research on Multicultural Education* 2nd ed. (San Francisco: John Wiley & Sons, 2004).
6. See Caroline Eick and Linda Valli, "Teachers as Cultural Mediators: A Comparison of the Accountability Era to the Assimilation Era," *Critical Inquiry in Language Studies* 7 (September 2010).
7. Refer to Anderson and Byrne, *The Unfinished Agenda of Brown v. Board of Education*, earlier cited.
8. Arthur B. Kennickell, "A Rolling Tide: Changes in the Distribution of Wealth in the U.S., 1989–2001," Levy Economics Institute, Table 10 (November 2003), http://www.faireconomy.org/research/wealth_charts.html (January, 2006).
9. Gary Orfield, "*Schools More Separate: Consequences of a Decade of Resegregation,*" *Harvard University, Civil Rights Project website,* http://www.civilrightsproject.harvard.edu/research/deseg/separate_schools01.php (January 2006). See also Peter Irons, *Jim Crow's Children: The Broken Promise of the Brown Decision* (New York: Viking, published by Penguin Group, 2002).
10. See U.S. Hate Crimes: Definitions and Facts, http://www.religioustolerance.org/hom_hat3.htm (January 2006).
11. Although the military has been considered, especially with the recent inclusion of women, as our most integrated institution since World War II, it is, except for drafts (which ended with the military withdrawal from Vietnam in 1973), an all-voluntary military.
12. See Patrick J. Ryan, "A Case Study in the Cultural Origins of a Superpower: Liberal Individualism, American Nationalism, and the Rise of High School Life, A Study of Cleveland's Central and East Technical Schools, 1890–1918," *History of Education Quarterly* 45 (Spring 2005): 66–95.
13. Ibid.
14. "Massification" refers to the expansion of mass education as part of modernity. See John Boli, Francisco O. Ramirez, and John W. Meyer, "Explaining the Origins of Expansion of Mass Education," *Comparative Education Review* 29 (1985): 145–170. Also, Marginson explains how educational aims will increasingly focus on developing capacities for adaptation to new environments and foreign associations. He contrasts this latest development to education of the Enlightenment, which emphasized strong links between family, work, and citizenship. See Simon Marginson, "After Globalization, Emerging Politics of Education," *Journal of Education Policy* 14 (1999): 19–35.

15. Benjamin Barber quoted in R.C. Salomone, *Vision of Schooling: Conscience, Community, and Common Education* (New Haven: Yale, 2000), 11.
16. Howard Zinn, *A People's History of the United States* (New York: Harper Collins, 2003), 11.
17. Dynamics of acculturation and minority students' many responses to forces of assimilation, from resistance to accommodation, have been widely studied by educational ethnographers. For example, see works by Michelle Fine, *Framing Dropouts* (Albany: State University of New York Press, 1991); Anne Locke Davidson, *Making and Molding Identity in Schools* (New York: State University of New York Press, 1996); Jeremy Price, *Against the Odds* (Stamford, CT: Alex Publishing Corporation, 2000); by anthropologists, including John Ogbu, "Immigrant and Involuntary Minorities: A Cultural-Ecological Theory of School Performance with Some Implications for Education," *Anthropology and Education Quarterly* 29 (1998): 155–188; and by historians of education such as V.P. Franklin, "Black High School Activism in the Late 1960s: An Urban Phenomenon?" *Journal of Research in Education* 10 (2000): 3–8; Victoria-María MacDonald, *Latino Education in the U.S.: A Narrative History, 1531–2000* (2004).
18. Term used by Marginson. See Simon Marginson, "After Globalization, Emerging Politics of Education," earlier cited.
19. Howard Zinn, *A People's History of the United States*, 10.
20. Amy Gutmann, *Democratic Education*, 309.
21. Maxine Green quoted by Judith A. Ramaley, The Presidents/Fourth of July Declaration on the Civic Responsibility of Higher Education, http://www.compact.org/resources/plc-declaration.html (December 2005).

Introduction

1. I borrow from Fine, Weiss, and Powell to define integration as "intellectual and social engagement across racial and ethnic groups." See Michelle Fine, Lois Weiss, and Linda C. Powell, *Harvard Educational Review* 67 (Summer 1997): 248. Thus I contrast the notion of *integration* to that of *desegregation*, which refers foremost to admission policies and processes to secure laws that ensure equal representation of students across race in schools.
2. Amy Gutmann, *Democratic Education* (Princeton, NJ: Princeton University Press, 1999), 161. See also Amy Stuart Wells et al., "How Society Failed School Desegregation Policy: Looking Past the Schools to Understand Them," *Review of Research in Education* 28 (2005): 47–99. Wells and her coauthors report how "recorded attitudes had changed relatively dramatically in the 40 years since the Brown decision...the percentage of Americans of all races who believed that the Supreme Court was right in its *Brown* decision increased from 63% in the early 1960s to 87% in the mid-1990s...percentage of Americans who believed that more should be done to integrate schools had risen rapidly in short time span, from 37% in 1988 to 56% in 1994," p. 77.

The authors use the word "integration" rather than "desegregation" to refer to balanced representation of students across racial groups.
3. See Jeanie Oakes, *Keeping Track: How Schools Structure Inequality* (Yale University Press, 2006).
4. Schofield (1989) quoted in Amy Stuart et al., "How Society Failed School Desegregation Policy: Looking Past the Schools to Understand Them," *Review of Research in Education* 28 (2005): 71.
5. Jencks quoted in Amy Stuart Wells et al., "How Society Failed School Desegregation Policy: Looking Past the Schools to Understand Them," *Review of Research in Education* 28 (2005): 71.
6. Gerald Grant, *The World We Created at Hamilton High* (Cambridge, MA: Harvard University Press, 1988). This analysis also builds closely on the work of William Graebner, *Coming of Age in Buffalo: Youth and Authority in the Postwar Era* (Philadelphia: Temple University Press, 1990). With the works of Grant and Graebner, this history shares its focus on youth intergroup relations across race divides (Grant) and across race and class and gender divides (Graebner). Unlike Graebner's work, but similar to Grant's work, it situates its youth within school boundaries. However, while both Grant and Graebner examine youth in cities (Grant's city counted 220,000 people in 1953 compared to Miller Town's 2,077 in 1950—and Graebner's city was Buffalo), this analysis focuses on a third-tier suburban town that develops over time from rural-suburban to urban-suburban, and it examines more closely how peer relationships are historically shaped across markers of difference within the extracurricular and curricular spaces of school. More important, it identifies students' situated positions within intersections of gender, class, race, religion, and nationality. To my knowledge this is the only other history of education that examines youth intergroup relationships, and the only history of education that examines young people's experiences from multiple intersecting categories of identity. In general, cultural historians and historians of youth have submerged students' experiences under the weight of economic, demographic, and political analyses. They have preferred to situate young people outside the perimeters of high school, and have represented them one-dimensionally. See John Gillis, *Youth and History: Tradition and Change in European Age Relations 1779–Present* (NY and London: Academic Press, 1974); Joseph Kett, *Rites of Passage: Adolescence in America, 1790 to the Present* (NY: Basic Books, 1977); Stephen Shlossman, *Love and the American Delinquent* (Chicago: University of Chicago Press, 1977); John Gilbert, *Cycles of Outrage: America's Reaction to the Juvenile Delinquent in the 1950s* (NY: Oxford University Press, 1986); R. Cohen, "The Delinquents: Censorship and Youth Culture in Recent U.S. History," *History of Education Quarterly* 37 (1997): 251–270; John Modell, *Into One's Own: From Youth to Adulthood in the United States, 1920–1975* (Berkeley: University of California Press, 1989); Grace Palladino, *Teenagers: An American History* (Chicago: University of Chicago Press, 1996); Joe Austin and Michael Willard, eds., *Generations of Youth: Youth Culture and History in Twentieth-Century America* (NY: New York University Press, 1998). The few histories that do situate youth within the high school setting and that attempt

to address gender, class, and racial differences do not reveal interactions between students across gender, race, and class boundaries. See E.F. Frazier, *Negro Youth at the Crossroads: Their Personality Development in the Middle States* (NY: Scholar Books, 1967); Kenneth Fish, *Conflict and Dissent in the High School* (NY: Bruce Publishing Co., 1970); Paula Fass, "Creating New Identities: Youth and Ethnicity in New York City High Schools in the 1930s and 1940s," in *Generations of Youth: Youth Culture and History in Twentieth Century America,* Joe Austin and Michael Willard, eds. (NY: New York Press, 1998), 95–117.

7. See addendum "Methodology: The Transparent Historian" for an in-depth discussion of oral histories as social acts and the historical consciousness.

8. Most recent works on high school activism include Dwayne C. Wright, "Black Pride Day, 1968: High School Student Activism in York, Pennsylvania," *The Journal of African-American History* 88 (Spring 2003): 151–162; Dionne Danns, "Chicago High School Students' Movement for Quality Education, 1966–1971," *The Journal of African-American History* 88 (Spring 2003): 138–150. See also V.P. Franklin, "Black High School Student Activism in the 1960s: An Urban Phenomenon?" *Journal of Research in Education* 10 (Fall 2000): 3–8. For an overview of works on African-American student activism in colleges as well as secondary institutions, see V.P. Franklin, "Introduction: African American Student Activism in the 20th Century," *The Journal of African-American History* 88 (Spring 2003): 105–109.

9. See Robert Palumbos, "Student Involvement in the Baltimore Civil Rights Movement, 1953–63," *Maryland Historical Magazine* 94 (1999): 449 and 485.

10. Information gathered from discussions with Howell S. Baum, who is presently working on a book entitled *How Liberalism Failed Brown: School Desegregation in Baltimore.* Howell Baum is professor of urban studies and planning at the University of Maryland, College Park. See also Steven Hahn, *A Nation Under Feet: Black Political Struggles in the Rural South from Slavery to the Great Migration* (Harvard University Press, 2004). In this award-winning history, Hahn brings to light the role of biracial alliances in African Americans' quest for self-governance, alliances that echo the African American and white alliances addressed in this history, whether in Baltimore prior to desegregation or in Miller Town following desegregation.

11. Thomas J. Vicino, *Transforming Race and Class in Suburbia: Decline in Metropolitan Baltimore* (NY: Palgrave Macmillan, 2008), 42. Between 1950 and 1970, the total population of Miller Town rose from 2,077 to 14,037. In 1950, whites represented approximately 92 percent of the population, and blacks were roughly 8 percent. In 1970, as white flight intensified, the percentage of whites rose to 97 percent of the population (from 1,912 registered white citizens in 1950 to 13,724 in 1970) compared to the black population, which, while it decreased in proportion to approximately 3 percent of the total population, increased from 165 registered African Americans in 1950 to 313 in 1970. See Table 3 for census data from the U.S. Department of Commerce Bureau of the Census.

12. See works by Louis S. Diggs: *Since the Beginning: African American Communities in Towson* (Baltimore: Uptown Press, 2000); *Holding on to Their Heritage* (Catonsville, MD: privately printed, 1996); *In Our Own Voices: A Folk History in Legacy* (Catonsville, MD: privately printed, 1998).
13. For discussions of changing immigration patterns since the Immigration Act of 1965 and implications for school populations, see works by Laurie Olsen, *Made in America: Immigrant Students in Our Public Schools*. (NY: New Press, 1997); and Alejandro Portes and Rubén Rumbaut, *Immigrant America: A Portrait*, 2nd ed. (Berkeley: University of California Press, 1996).
14. Gerald Grant, *The World We Created at Hamilton High*, 6.
15. For elaborations on multivocal narratives, see Peter Burke, *History and Social Theory* (NY: Cornell University Press, 2005). For ethnographies that examine students' high school experiences across markers of difference, see August Hollingstead, *Elmtown's Youth: The Impact of Social Class and Adolescents* (NY: J. Wiley, 1949); Penelope Eckert, *Jocks & Burnouts: Social Categories and Identity in the High School* (NY: Teachers College Press, 1989); Michelle Fine, *Framing Dropouts* (Albany: State University of New York Press, 1991); Ellen Brantlinger, *The Politics of Class in Secondary School* (NY: Teachers College Press, 1993); Jay MacLeod, *Ain't No Makin' It* (Boulder, CO: Westview Press, 1995); Ann Locke-Davidson, *Making and Molding Identity in Schools* (NY: State University of New York Press, 1996); Laurie Olsen, *Made in America: Immigrant Students in Our Public Schools* (NY: New Press, 1997).
16. See Patricia Collins, *Black Feminist Thought: Knowledge, Consciousness, and the Politics of Empowerment* (NY: Routledge, 2000); Patricia Collins, *Fighting Words: Black Women and the Search for Justice* (Minneapolis: University of Minnesota Press, 1998).
17. See Kimberlé Crenshaw, Neil Gotanda, Gary Peller, and Kendall Thomas, eds., *Critical Race Theory: The Key Writings that Formed the Movement*, (NY: New York Press, 1995); Bill Ashcroft, Gareth Griffiths, and Helen Tiffin, *Post-Colonial Studies: The Key Concepts* (NY: Routledge, 2005); as well as the works of Patricia Collins earlier cited.
18. Leslie McCall, "The Complexity of Intersectionality," in *Signs: Journal of Women in Culture and Society* 30 (2005), http://www.journals.uchicago.edu/doi/abs/10.1086/426800.
19. Alessandro Portelli, "What Makes Oral History Different," in *The Oral History Reader*, ed. Robert Perks and Alistair Thomson (NY: Routledge, 2005), 63–74 (first emphasis mine, second emphasis that of the author). See also Alessandro Portelli, *The Death of Luigi Trastulli and Other Stories: Form and Meaning in Oral History* (Albany: SUNY Press, 1990).
20. Thirty-seven alumni of different races and socioeconomic backgrounds, identified through yearbooks and alumni directories as well as through snowball sampling, and who graduated from Miller High between 1954 and 2002, were formally interviewed in tape-recorded sessions that ranged from one to two hours, and informally through follow-up calls, visits, and e-mail communications. These alumni were chosen for their differences (gender and class just prior to 1956 desegregation; gender, class, and race following

desegregation), and situated as males and females of different racial and economic backgrounds, across the decades, as much as possible. The decade boundaries emerged from changes in oral testimonies as well as corroborating changes examined in yearbooks, community newspapers, census data, and other documentary sources. See Addendum: Methodology.
21. In examining yearbooks, total populations of students and distribution of student population by race and gender were tabulated, as well as distribution of students across race and gender in extracurricular activities and clubs. For descriptions of archival research, see Addendum: Methodology.
22. Irma Omeldo, "Redefining Culture through the Memorias of Elderly Latinas," *Qualitative Inquiry* 5 (1993): 353–376.
23. See Roger Brown, *Social Psychology* (NY: Free Press, 1986); see also Robert Levine and Donald Campbell, *Ethnocentrism: Theories of Conflict, Ethnic Attitudes, and Group Behavior* (NY: John Wiley & Sons, 1972).
24. See Roger, J.R. Levesque, *Dangerous Adolescents, Model Adolescents: Shaping the Role and Promise of Education* (NY: Kluwer Academic/Plenum Publishers, 2002), 53–54. In discussing the ruling in *Tinker vs. Des Moines Independent Community School District*, 1969, Levesque explains, "[S]tudents challenged a school's prohibition against wearing black arm bands in protest of the Vietnam War." Students won as it was judged that "minors were persons protected by the Constitution."
25. Zero-tolerance policies, which spread in the mid-1990s, targeted possession or use of drugs or alcohol and weapons by students on school premises. In some cases, zero-tolerance policies became opportunities for school authorities to threaten youth beyond evidence of wrongdoing. For opinions in support of zero-tolerance policies, see National Safety Guard and Security Services, Zero Tolerance, http:www.schoolsecurity.org/trends/zero_tolerance.html. For critiques, see Dennis Cauchon, "Zero Tolerance Policies' Lack of Flexibility," *USA Today*, April 13, 1999.
26. See Tyack and Cuban, *Tinkering Toward Utopia: A Century of Public School Reform* (Cambridge, MA: Harvard University Press, 1995), 44–45. Tyack and Cuban remark about the politically conservative periods of the 1950s and 1980s: "Policy talk about schools stressed a struggle for national survival and international competition...In such periods, policy elites want to challenge the talented, stress the academic basics and press for greater coherence and discipline in education." The authors suggest that "liberal eras such as the 1930s and 1960s stress ideology of access and equality." Tyack and Cuban remark, however, that by the close of the twentieth century, in the "late 80s and 90s...conservatives and liberals alike...called for national standards." By the end of the twentieth century, cyclical, pendulum-like criticisms of high school performance reach the highest level of attack on the legitimacy of the high school within the 1950–2000 period. See also work by William Wraga, *Democracy's High School: The Comprehensive High School and Educational Reform in the United States* (Lanham, MD: University Press of America, 1994).
27. See Grace Palladino, *Teenagers: An American History* (NY: Basic Books, 1996).

28. See report by Gary Orfield and Chungmei Lee, "Brown at Fifty: King's Dream or Plessy's Nightmare," Civil Rights Project Harvard University, http://www.civilrightsproject.ucla.edu/research/reseg04/resegregation04.php (accessed June 24, 2010).
29. See Clayborne Carson, "Two Cheers for Brown v. Board of Education," *Journal of American History* 91 (June 2004), 26–31.
30. Carnoy and Levin, quoted in David F. Labaree, *How to Succeed in School Without Really Learning* (New Haven: Yale University Press, 1997), 49.
31. Wells et al., "How Society Failed School Desegregation Policy: Looking Past the Schools to Understand Them," *Review of Research in Education*, 28 (2005): 90; "[C]lose friendships within racially diverse schools are more often than not same-race friendships...." This historical analysis shows, however, that interracial friendships can flourish under particular social and demographic conditions such as those experienced by the Border-Crossing Generation.
32. Findings by Amy Stuart Wells et al. also suggest "that there are several ways to interpret lunchroom segregation within desegregated schools...and one is to consider who students see and interact with outside of school" (Wells, 2005, p. 90). Certainly there are indications that even at the beginning of desegregation at Miller High, during the Divided Generation interracial playmates of Miller Town were more likely to sit together in the cafeteria during lunchtime. Refer to Annie Cole's testimonies. I suggest that beyond imported habits of association, students also segregated in the cafeteria at the end of the century because of the absence of adult role-modeling and adult engagement.
33. For a discussion of the nefarious role of tracking on student achievement, see Jeanie Oakes, *Keeping Track: How Schools Structure Inequality* (Yale University Press, 2006).

1 Memories of Class, Race, and Gender Divisions: Immediate Pre- and Post-Desegregation Years (1950–1969)

1. The name of the school and all people's names have been replaced with pseudonyms to protect those graduates unwilling to be identified by testimonies they felt might compromise their professional and familial relationships in the community. Some narrators have agreed to reconsider release of transcripts at a later date. See Addendum: Methodology.
2. <Miller> High was first known as <Miller> Academy, founded in 1820. It was renamed <Miller Town> High School in the 1840s, and then renamed <Miller> High in the 1950s. For details about the various site and building transformations of the high school, I referred to C.P., *<Miller's> Century of Progress 1878–1978* (<Miller Town>: C.P., 1978), located in the <Miller Town> Library archives.

3. <Miller High> *Alumni Directory* (NY: Bernard C. Harris Publishing, 2000), ix. The 1930s building is further described as housing "the widest variety of educational equipment ever seen in this section, making it possible for the school to offer courses of instruction in practically every phase of academic, cultural, commercial, and industrial secondary education."
4. Sherry Parson (1958–1962), interview with the author, March 7, 2003.
5. C.P., *<Miller's> Century of Progress: 1878–1978* (<Miller Town>: C.P., 1978).
6. Sherry Parson (1958–1962), interview with the author, March 7, 2003. The bell from the original academy, built in the early nineteenth century, was preserved and kept in the high school throughout its transmutations. It is presently in a small garden and is still rung at the beginning of every school year.
7. From conversations with Annie Milligan (see Addendum: Methodology). See also Louis S. Diggs, *Since the Beginning: African American Communities in Towson* (Baltimore: Uptown Press, 2000); *Holding on to Their Heritage* (Catonsville, MD: privately printed, 1996); *In Our Own Voices: A Folk History in Legacy* (Catonsville, MD: privately printed, 1998). See also E. Franklin Frazier, *Black Bourgeoisie* (NY: Free Press Paperbacks, 1997). Frazier's work, originally published in 1967 and widely criticized for its harsh portrayal of middle-class blacks in the United States, problematizes the meaning of "middle class" for African Americans whose middle-class status did not de facto change their relative economic standing in the United States. Frazier's findings regarding conspicuous consumption among the African American middle class do not reflect the black community of Miller Town in this history. However, Frazier's seminal work provides a context for understanding the economic place of African American business owners, and certainly those of the 1950 to 1969 time period discussed in this chapter.
8. Linda Eisenman, *Higher Education for Women in Postwar America, 1945–1965: Reclaiming the Incidental Student* (Baltimore: Johns Hopkins University Press, 2006), 19. See also Kenneth. T. Jackson, *Crabgrass Frontier: The Suburbanization of the United States* (Oxford, NY, Toronto: Oxford University Press, 1985).
9. Linda Moss (1965–1969), interview with the author, September 24, 2003.
10. At that time, high school graduates could still be teachers. Kaufman (1950–1954), interview with the author, February 4, 2003: "I could get a job in teaching in the County without a degree."
11. Plays were performed regularly not only by students, but also by faculty and community members. Annual fairs were held where farmers showed livestock and sold produce. It was also an opportunity for the townspeople to attend free sporting matches, art shows, and live music.
12. To get a sense of the school climate at Miller High in 1956, I interviewed students who graduated in 1954 and 1956. These graduates' experiences covered the full first half of the 1950s.
13. Dorothy Kaufman (1950–1954), interview with the author, February 4, 2003.

14. Dorothy Kauffman (1950–1954), interview with author, February 4, 2003.
15. Yearbooks consistently reveal a majority of white women in honor society photographs. See Addendum: Methodology.
16. Robert Heart (1952–1956), interview with the author, June 24, 2006: "They expected less of the boys...[the teachers] weren't going to get anything out of [the boys]." Judy Law (1950–1954), interview with the author, February 25, 2003: "Girls were considered to be smarter." Alice Web (1950–1954), interview with the author, October 2, 2003: "Girls were the better students." See also James B. Conant, *The American High School Today: A First Report to Interested Citizens* (NY: McGraw-Hill, 1959), and his second report, *The Comprehensive High School: A Second Report* (1967). Although there is no possibility of assessing how many girls did homework for the boys, the fact that male and female alumni who do not know each other reported as much, and that they reported as much across time periods (within the broader study) justifies documenting this recollection. Future research might focus on recovering changes and continuities in female-male student relationships not in terms of achievement gaps, but in terms of study habits and the role that sexual interest plays (or doesn't play) in shaping them.
17. Robert Heart (1952–1956), interview with the author, June 24, 2006. Tim Whittle, alumnus who attended in the late 1970s, also recalled girls doing boys' homework (see part II, chapter 3).
18. Alice Web (1950–1954), interview with the author, October 2, 2003.
19. Ibid.
20. Yearbooks and alumni stories reveal a consistent majority of young women yearbook editors, writers, and illustrators, as well as school newspaper editors and writers. See Addendum: Methodology.
21. Linda Moss (1965–1969), interview with the author, September 24, 2003.
22. Sandy Eycke (1950–1954), interview with the author, March 18, 2003: "A large portion of my class started first grade...went through all twelve grades...everybody went to Miller at some point in my family...You had teachers who had taught most of us. A lot of teachers lived in the community...A lot of them knew parents, brothers, and sisters." Betty Land (1964–1968), interview with the author, February 8, 2003: "My mom is a Miller High graduate, as is her mom and dad and parents. So I have a long line of Miller graduates in my family. I have two children of my own that have graduated from Miller High, and I have one there now and one in the middle school." Linda Moss (1965–1969), interview with the author, September 24, 2003: "They were role models...the relationships were great...and it was so relaxed but firm, and everybody did what they were supposed to do."
23. <*The Key*> yearbook (1958), 74.
24. Alice Web (1950–1954), interview with the author, October 2, 2003.
25. Ibid.
26. Sandy Eycke (1950–1954), interview with the author, March 18, 2003: "I probably would have gone to higher education if we could have afforded it. I think that was by default...if you weren't going to college, this is what

was left." Some, like Nora Jones, who was artistically inclined, completed their studies within the general track and availed themselves of an "absolutely wonderful art program...[and] great art teachers." Quote from transcript of audiotape interview with Nora Jones (1955–59), interview with the author, June 12, 2003.
27. Judy Law, interview with the author, February 25, 2003.
28. Judy Law (1950–1954), interview with the author, February 25, 2003. Also, Alice Web, ((1950–1954), interview with the author, October 2, 2003), in remembering Judy Law, said, "She didn't go to college...and she was one of the brightest in the class. She never thought she was going to go to college...we were, to a certain extent, in our slots."
29. Graduates' recollections echo historian Beth Bailey's findings regarding the social practice of dating and "going steady." See Beth Bailey, *From Front Porch to Back Seat: Courtship in Twentieth-Century America* (Baltimore: Johns Hopkins University Press, 1998).
30. Judy Law (1950–1954), interview with the author, February 25, 2003: "The biggest negative for me at high school is that it didn't really prepare me for the real world...it didn't prepare me to question authority."
31. Dorothy Kauffman (1950–1954), interview with the author, February 4, 2003.
32. Judy Law (1950–1954), interview with the author, February 25, 2003.
33. Nora Jones (1955–1959) and Lou-Anne Kensington (1953–1957), interview with the author, June 12, 2003.
34. Nat Right (1957–1963), interview with the author, July 19, 2004.
35. Bud Land (1964–1968), interview with the author, August 16, 2003.
36. Dorothy Kaufman (1950–1954), interview with the author, February 4, 2003.
37. Alice Web (1950–1954), interview with the author, October 2, 2003: "The English teachers that I had were really, really good, and social studies...but the math was just rote...math and science was not well taught...I would have been really interested in biology...I had Mr. B., whom everybody loved, but he was not a good teacher."
38. Judy Law (1950–1954), interview with the author, February 25, 2003: "For me it was a saving grace to be able to come to school. I got to dress up and came to school all pretty and clean in something my grandmother had made me...I traveled. I mean, it was travel from the country to the city...Oh, it was a big social time."
39. Judy Law (1950–1954), interview with the author, February 25, 2003: "[The diploma] meant that I had a job. And that I was free...to leave home. That was what it meant to me. My freedom. When they think of freedom, most people think of it as a patriotic thing. I think of it as a very personal thing."
40. Ibid. "It was a big deal that I graduated...I think it gave them an elevated level of prestige. My father wanted me to graduate because he never had; my mother wanted me to graduate because it was her school and because, I guess, it reflected on her in some way."

41. Dorothy Kaufman (1950–1954), interview with the author, February 4, 2003.
42. Alice Web (1950–1954), interview with the author, October 2, 2003.
43. Sherry Parson (1958–1962), interview with the author, March 7, 2003.
44. Linda Moss (1965–1969), interview with the author, September 24, 2003.
45. Dorothy Kaufman (1950–1954), interview with the author, February 4, 2003.
46. Ibid. Also, Alice Web (1950–1954), interview with the author, October 2, 2003: "Every time you had a class, she'd rearrange the class, and the person that got the best grade on the test would sit in the first seat in the first row. And then it would go all the way to the person that got the worst grade in the back of the room...I guess that shows you the advantage that the bright kids had, and very often the bright kids were also the ones who had more advantages at home, too. And got more support at home."
47. Betty Land (1964–1968), interview with the author, February 8, 2003: "Every Friday we had basketball, boys' basketball, and that was really a big thing then. And everybody would go to the games...Big school spirit. And we always had the cheerleaders. We always had a girl that was dressed up like...." Linda Moss (1965–1969), interview with the author, September 24, 2003: "We had great plays. We had great pep rallies."
48. Alice Weber (1950–1954), interview with the author, October 2, 2003: "And we all knew the whole story of graduation and we all just loved it, you know, the pomp and circumstance of it. We sang 'I'll never walk alone'...it was a tradition...I knew it was a place I was really tied to, it was part of my life."
49. Dorothy Kaufman (1950–1954), interview with the author, February 4, 2003: "I do remember one girl who...dropped out. Because her parents insisted that she [girl with learning disability] be allowed to come to school and to, you know, have the social experience. She was a sweetheart. But she was so lonely; you could tell she was lonely because she wouldn't come join in unless you went to get her. You know, she didn't have the confidence."
50. Alice Web (1950–1954), interview with the author, October 2, 2003: "There was one girl who was pregnant when we graduated, but she was wild, you know...everybody knew she was wild...She was like four or five months pregnant." Dorothy Kaufman (1950–1954), interview with the author, February 4, 2003: "One pregnant girl...just disappeared." Linda Moss (1965–1969), interview with the author, September 24, 2003: "There were pregnant girls in our class. Was pretty scandalous."
51. Alice Web (1950–1954), interview with the author, October 2, 2003.
52. Ibid.
53. Robert Heart (1952–1956), interview with the author, June 24, 2003.
54. Ibid.
55. Robert Heart's experiences as a poor farm boy echoes those of poor white sharecropping families recorded by the writer James Agee and the photographer Walker Evans in *Let Us Now Praise Famous Men: Three Tenant Families* (Boston: Houghton Mifflin/Mariner, 2001), originally published in 1941.

56. Robert Heart (1952–1956), interview with the author, June 24, 2003: "I would go on my paper route from one to three o'clock in the morning. Come home, grab a couple of hours sleep, then milk six cows from 5:30 to 6:30, then go to school. Get off school. Go home, milk six cows and then work for Shellborn in Miller Town at the gas station, from seven to eleven at night…I did that every day. And then, when the weekends came around, I would help on the weekend paper. That's what I did. That's how I got through high school…I was always sleepy.…Most of the farm boys didn't have time to do their homework…They knew we didn't do it…We only got to learn what's in the class."
57. Robert Heart (1952–1956), interview with the author, June 24, 2003.
58. There was also a sexual edge to these boys for which, as "red blooded American boys," they expected to be understood. Robert Heart (1952–1956), interview with the author, June 24, 2003: "I got kicked out of one my English classes and I got sent to the principal's. And he asked me, 'What is going on in that classroom for you to be kicked out?' I said, 'All of the boys are sitting in the front of the classroom…she wanted us to sit in the front of the classroom. And she sits on the desk swinging her legs. Now, none of us could concentrate on anything.' So he walked down the hall. She was sitting on the desk with the boys in front. Well, after that I had no problems with her. I still got a D…They liked to pick on the farm boys…their dads weren't educated."
59. Robert Heart (1952–1956), interview with the author, June 24, 2003: "One of my favorite teachers was the shop teacher…He covered for us all the time…He got me through high school…Instead of spending two hours in detention, he would say, 'Come down here I need you. I need this work done. I need you to draw the floor plan.'…So I would go down to the shop class, sit at the drawing table, and draw…He got me pretty much through high school…He'd say, 'I'd love to grab you by the ear and say this is what you have to do to get out of here.' He helped me through a lot."
60. Ibid.
61. Ibid.
62. Annie Cole (1957–1959), interview with the author, March 11, 2003.
63. Robert Heart (1952–1956), interview with the author, June 24, 2003.
64. Ibid.
65. Sherry Parson (1958–1962), interview with the author, March 7, 2003.
66. Ibid.
67. Robert Heart shared how he would try mimicking the ways of those with money. He was acutely aware of being shunned by people whom he called "upper class," and he tried to date "rich" girls who continually refused him. It is also essential to point out that while farm boys across time periods were remembered as "hillbillies" and "rednecks" and reviled by alumni as racist, there was no evidence I could find in Robert Heart's recollections to suggest that he was a racist farm boy. It is equally important to note that Robert graduated in the summer of 1956, a couple of months before desegregation.
68. Ibid.

69. Robert Heart (1952–1956), interview with the author, June 24, 2003. Descriptions of high school as a prison-like or regimented place were echoed by other white lower-class male students across time periods, in particular Tim Whittle, a 1981 graduate (see part II, chapter 3).
70. Robert Heart (1952–1956), interview with the author, June 24, 2003.
71. Ibid.
72. Robert Heart (1952–1956), interview with the author, June 24, 2003: "If you made a mistake at the beginning of the first problem, you failed it...and then you would go back and do the problem, but he wouldn't do the problem for the speed of the slowest. He would do the problem for the speed of the fastest...So I failed his class...So I switched to Mrs. Hill, and she knew where I was coming from, okay [fatherless farm boy on welfare]...and then all of a sudden it makes sense...that's what she did and I got through it with no problems." This memory illustrates the allowances Robert expected precisely because of his life circumstances.
73. Robert Heart (1952–1956), interview with the author, June 24, 2003: "[We were] telling the teachers, look, we're going to school. Care about us. We may not be paying attention, but we are going to school."
74. Robert Heart (1952–1956), interview with the author, June 24, 2003.
75. Sherry Parson (1958–1962), interview with the author, March 7, 2003.
76. Robert Heart (1952–1956), interview with the author, June 24, 2003.
77. Bud Land (1964–1968), interview with the author, August 16, 2003: "I had had it [with high school]. Literature? I don't want to use it at all. I don't want to deal with it. Where is it going to fall into my life?"
78. Bud Land (1964–1968), interview with the author, August 16, 2003.
79. Ibid.: "Betty went along with the very conforming group....You can look at the yearbooks and see Betty and her friends hanging around in ninth grade, and it's the same group in twelfth grade." Betty, as earlier discussed, was one of the white middle-class girls who understood themselves to be, as Linda Moss expressed, "the doers."
80. Bud Land (1964–1968), interview with the author, August 16, 2003.
81. Ibid.
82. Bud Land (1964–1968), interview with the author, August 16, 2003.
83. Doris Right (1956–1958), interview with the author, September 4, 2003.
84. Annie Cole (1956–1958), interview with the author, March 11, 2003.
85. Ibid.
86. Doris Right (1956–1958), interview with the author, September 4, 2003.
87. Annie Cole (1956–1958), interview with the author, March 11, 2003.
88. Ibid.
89. Ibid.
90. Doris Right (1956–1958), interview with the author, September 4, 2003.
91. Ibid.: "My experiences with the teachers were great. As far as I knew them, I had no problems with the teachers, they appeared very helpful, especially Miss R., who I understand is deceased now. [Then there was] Miss T., was her name. She was my speech teacher...she made me talk...she was a sweet lady, she really was." Annie Cole (1956–57), interview with the author, March

11, 2003: "I was a good baseball player. And the gym teachers liked me a lot...Mrs. Streesby, she was a very nice teacher. She was one of the teachers that you could go and talk to. And she didn't take sides with anybody. She didn't like nobody to be called names, and you did get in trouble if you called names and she heard you. She was good like that...But it was a few of them."

92. Annie Cole (1956–1957), interview with the author, March 11, 2003.
93. Dorothy Kaufman (1950–1954), interview with the author, February 4, 2003.
94. Annie Cole (1956–1957), interview with the author, March 11, 2003.
95. Doris Right (1956–1958), interview with the author, September 4, 2003.
96. Annie Cole (1956–1957), interview with the author, March 11, 2003.
97. Doris Right (1956–1958), interview with the author, September 4, 2003.
98. Annie Cole (1956–1957), interview with the author, March 11, 2003 .
99. Ibid: "You know, the parents, the white parents had been there for years. Some of the families been in the community helping the school. It wasn't much he could do. He tried. He made things as comfortable as he could for us. I give him that."
100. Annie Cole (1956–57), interview with the author, March 11, 2003: "I'll tell you what made me really, really feel good. When I was sitting in my classroom one day and Mr. Lancaster called me over the intercom. Told the teacher, 'Could you please send Miss Cole down to the office.' I said, 'Oh my God, what have I done?' I was trying to remember what I did wrong...He said, 'Come in, Miss Cole...have a seat and relax and just have a seat and sit down.' I said okay. And so he says to me, 'I'm very proud of you.' I said, 'Why? What did I do?' He said, 'You're very bright and you're doing a very beautiful job...I'm going to put you on the honor roll...You keep up the good work. I know it's hard, but you're doing a wonderful job. You keep doing what you are doing.' So I remember that. That stuck with me."
101. Annie Cole (1956–1957), interview with the author, March 11, 2003.
102. Biographies accompanying senior pictures were examined in yearbooks for "tracking."
103. Judy Law (1950–1954), interview with the author, February 25, 2003: "Because I lived outside of town I had to depend on the bus to take me back and forth...it was very difficult for me to do anything after school and I usually didn't.... We had one car and my father worked seven days a week."
104. Annie Cole (1956–1957), interview with the author, March 11, 2003: "I worked after school. And I used to leave from school and walk to Emory Grove, and I used to clean the cottages over there...a couple of days a week...I was tired when I was finished because I had to walk back home...it's quite a walk. The lady used to give me eight dollars, and I used to give four dollars to my mom to put up and she let me have four dollars for my pocket. And at the end of the year, she'd take me downtown shopping." Doris Right (1956–1958), interview with the author, September 4, 2003: "We were very sheltered...we had to go home immediately after school...because my mother

had to work at night, and she wanted to make sure that we were home. And my father was a truck driver, so he wasn't home that much...She made sure we were in bed—dinner, homework, and bed."
105. This finding echoes educational ethnographer Michelle Fine's findings in the late 1980s of minority students dropping out proactively out of self-preservation, rather than out of failure. See Michelle Fine, *Framing Dropouts*.
106. Annie Cole (1956–1957), interview with the author, March 11, 2003: "Everybody would kind of stay away from the blacks...I couldn't, I just couldn't, I left....I just felt out of place at all times. I just didn't like it. My brother [one year older] liked it. My sister [four years younger] liked it...I graduated...from home."
107. Doris Right (1956–1958), interview with the author, September 4, 2003. Doris's memory of the details of the difference in curriculum between Miller and the all-black high school remained vague, while her emotional memory of the "different things" taught was strongly expressed in tone and inflection.
108. Doris Right (1956–1958), interview with the author, September 4, 2003: "All five of us [from the class of 1958] graduated."
109. Doris Right (1956–1958), interview with the author, September 4, 2003.
110. Norman Good (1956–1959), interview with the author, March 18, 2003.
111. Ibid.
112. Annie Cole (1956–1958), interview with the author, March 11, 2003: "My brother, he was the track star...he did a lot of things for the school. He was an honor roll student. He was just a bright kid. He really got along fine with everybody."
113. Norman Good (1956–1959), interview with the author, March 18, 2003.
114. Willie James and Jimmy Cole are deceased. Norman Good and Annie remembered them.
115. Football was considered too dangerous a sport in Baltimore County until the end of the 1960s, when it was first introduced.
116. Norman Good (1956–1959), interview with the author, March 18, 2003.
117. Ibid.
118. Norman Good (1956–1959), interview with the author, March 18, 2003: "[O]ne good friend, for the record, Frank Martell, we were on the track team together. He was white. We developed an excellent bond...he was the only person I let call me Norm. You know, we had that kind of a rapport and all. And he was a very popular individual...everybody knew Frank, because he had that kind of personality, he was a people person and all."
119. Norman Good (1956–1959), interview with the author, March 18, 2003.
120. Ibid.
121. The question remains why capable black young women did not make it into the academic track since they too would not have been recommended for a white higher education establishment after graduation. Did African American young women lack the parental support that would set them on a course toward higher education? Further research is required.
122. Norman Good (1956–1959), interview with the author, March 18, 2003.
123. Doris Right (1956–1958), interview with the author, September 4, 2003.

124. Ibid.
125. Nat Right (1958–1963), interview with the author, July 19, 2004.
126. Nat Right (1958–1963), interview with the author, July 19, 2004. Doris Right (1956–1958), interview with the author, September 4, 2003: "The one thing that really, really upset our family, probably the entire community...my brother had stayed at my grandmother's house who lived in Smithville. And the people that she worked for, the Griffiths, brought him to school...Because at that time, I don't think my grandmother had a telephone...the message didn't get to him. So he was dropped off at school, he walked around the corner, and he [a white male adolescent] hit him in the mouth with brass knuckles and he [her brother] lost all his front teeth...that was the worst thing that ever happened."
127. Annie Cole (1956–1957), interview with the author, March 11, 2003. The brother in question who hit the white boy was not Annie's popular brother. Annie also recalled: "The principal got all the blacks together, and he was very stern and said there wouldn't be any trouble. But we tried to explain to him...and we told him that the teachers saw it and didn't stop it. He said...some things he couldn't control."
128. Annie Cole (1956–1957), interview with the author, March 11, 2003.
129. Dotty Morris (1963–1965), interview with the author, September 24, 2004.
130. Dotty Morris (1963–1965), interview with the author, September 24, 2003.
131. Ibid.
132. Dotty Morris (1963–1965), interview with the author, September 24, 2003: "I was involved and I liked to sing and so did my girlfriend, so we were always singing. We were in the glee club and the choir when I was in middle school. I was in music also there. [At Miller High] I was in the school orchestra. I played the violin."
133. Dotty Morris (1963–1965), interview with the author, September 24, 2003.
134. Dotty Morris (1963–1965), interview with the author, September 24, 2003: "Mary and Thomas were in my class."
   "Mary was white and Thomas was black?"
   "Yeah. And they're still married today, three kids."
   "Wasn't it unusual to see an interracial couple then?"
   "It was kind of the start, you know, at that time."
   "Were they shunned by people in the community?"
   "I don't really know about that, but I think that there were some, you know what I mean about that. But they [Mary and Thomas] continued on, and I don't know what it was that she heard, because, you know, I didn't go there."
135. Dotty Morris (1963–65), interview with the author, September 24, 2003.
136. Ibid.
137. Dotty Morris (1963–65), interview with the author, September 24, 2003.
138. Ibid.
139. Annie Cole (1956–1957), interview with the author, March 11, 2003: "My sister liked it [Miller High]...She's younger than me—four years."
   "Did she graduate or just attend?"
   "She graduated....She's a nurse."

140. Nat Right (1958–1963), interview with the author, July 19, 2004: "Jocks were popular."
    "Were you a jock?"
    "No. When I came through, there was James and Samuel. They were good sportsmen...They gravitated to the jocks, the whites and the black jocks."
141. Nat Right (1958–1963), interview with the author, July 19, 2004.
142. Ibid.
143. Nat Right (1958–1963), interview with the author, July 19, 2004.
144. Burt Sadden (1963–1967), interview with the author, July 26, 2004.
145. Ibid.
146. Burt Sadden (1963–1967), interview with the author, July 26, 2004: *"Did you participate in extracurricular activities?"*
    "No...it was go to school and work. After school it was always try to find part-time work, to help the family, and to survive...a few would stay, in the higher group, or the students, I should say, in the more intelligent classes seemed to hang around and do more at the school than just the general students. Most of the students did their hours and then got on the bus."
147. Burt Sadden (1963–1967), interview with the author, July 26, 2004.
148. Burt Sadden (1963–1967), interview with the author, July 26, 2004: "Blacks stayed to themselves and whites stayed to themselves."
    "Was that because of racial animosity?"
    "Pretty much culturally."
149. Burt Sadden (1963–1967), interview with the author, July 26, 2004.
150. Nat Right (1958–1963), interview with the author, July 19, 2004.

2 Cautiously Negotiating Social Divides: A Conservative Student Body (1950–1969)

1. Judy Law (1950–1954), interview with the author, February 25, 2003: "We were asked out by boys who were at another school...And I got asked out by a couple of people who were, I would say, out of my league...meaning their family had a lot of money...And I remember two in particular who asked my father to take me out and he said no...He just didn't like the idea. I remember telling my father later...you know, you may have just kept us from having the family fortune."
2. Alice Web (1950–1954), interview with the author, October 2, 2003.
3. Bud Land (1965–1969), interview with the author, August 16, 2003.
4. Alice Web (1950–1954), interview with the author, October 2, 2003.
5. Robert Heart (1952–1956), interview with the author, June 24, 2003: "We couldn't date the upper-class girls. They just wouldn't date us. They wouldn't. You would ask them for a date and they would just turn around and walk away. [But] if you were a junior [upper-class woman], you could date a senior of the lower class."

6. Robert Heart (1952–1956), interview with the author, June 24, 2003.
7. Norman Good (1956–1959), interview with the author, March 18, 2003.
8. Nat Right (1956–1963), interview with the author, July 19, 2004.
9. Doris Right (1956–1958), interview with the author, September 4, 2003: "My parents had bought land on the Corner of C. Avenue...." The Rights had moved to a wealthier white neighborhood. As discussed in the Introduction, black families of Miller Town of the 1950s and 1960s owned their own businesses; moreover, these families' roots reached back into the nineteenth century and their mobility within Miller Town would prove relatively easy given their long-established standing within the majority white community.
10. White and black, female and male graduates reported attending either the local Episcopal, Catholic, or Methodist churches.
11. Annie Cole (1956–1957), interview with the author, March 11, 2003: "They [white students] were throwing Ivory Soap on the floor...I thought they were maybe trying to make the floor slippery so we could dance better. You know, get more sliding and doing. But no, ma'am...it was to instigate a fight. I didn't understand what was really going on...I thought we [the black students] just want to dance...So it finally escalated. Somebody hit somebody. Somebody called somebody that bad name."
12. See Table 3 in Addendum, Methodology: The Transparent Historian. See also Thomas J. Vicino, *Transforming Race and Class in Suburbia: Decline in Metropolitan Baltimore*, 42.
13. Annie Cole (1956–1957), interview with the author, March 11, 2003: "They ran us out of the school." Teachers' tacit support of white students' rejection of blacks on the dance floor was most likely also a reflection of their fear of black expression. Historian Grace Palladino explains that "Rock 'n' roll was everything that middle-class parents feared: elemental, savage, dripping with sexuality, qualities that respectable society usually associated with depraved classes." See Grace Palladino, *Teenagers: An American History*, 155.
14. Linda Eisenman, *Higher Education for Women in Postwar America, 1945–1965: Reclaiming the Incidental Student* (Baltimore: Johns Hopkins University Press, 2006), 19.
15. Ibid, 18.
16. See Paula Fass, *The Damned and the Beautiful: American Youth in the 1920s* (NY: Oxford University Press, 1977); Barbara Finkelstein, "Is Adolescence Here to Stay? Historical Perspectives on Youth and Education," in *Adolescence and Society*, eds., T. Urban and F. Pajares (NY: Information Age Press, 2003), 1–33.
17. Eisenman, *Higher Education for Women in Postwar America, 1945–1965: Reclaiming the Incidental Student*, 18.
18. It is worth noting that none of the white female narrators reported aggressive behavior by males toward them or toward other white women they may have recalled. All alumni were asked to describe their perceptions of and relationships with the opposite sex at high school. More research is needed to see if white adolescent girls were in fact less likely than their black counterparts to be sexually harassed by white adolescent boys.

19. Doris Right (1956–1958), interview with the author, September 4, 2003.
20. Franklin Frazier, *Negro Youth at the Crossroads: Their Personality Development in Middle States* (NY: Scholar Books, 1967), 105. See also Barbara Finkelstein, cited earlier.
21. Annie Cole (1956–1957), interview with the author, March 11, 2003.
22. Annie Cole (1956–1957), interview with the author, March 11, 2003: "I played girls' softball, which was fun. You know, when you're playing sports, everybody [narrator smiled remembering good times], yeah, now that was fun." Nora Jones (1955–1959), interview with the author, June 12, 2003: "Rose [African American peer], who played hockey, you just didn't want to meet that girl coming down the hockey field [narrator smiled with appreciation for Rose's skills]."
23. Annie Cole (1956–1957) excelled in gym; Robert Heart loved drawing in shop classes, and after years of diverse occupations (from soldier in the army to working for an architectural firm), became an architect without ever having attended college; Nora Jones became an artist, and raved about all the art teachers, etc.
24. See works by Samule Bowles and Herbert Gintis, *Schooling in Capitalist America* (NY: Basic Books, 1976). Yet, against socially reproductive norms, testimonies often revealed the importance of relationships with teachers who cared for youth across race and class divides; relationships with teachers that changed or helped the lives of alumni—including African American Norman Good, who went on to hold prestigious positions in corporations and the federal government. He recounted having been able to land better jobs in mostly white settings because of the vote of confidence he had received from his math and foreign language teacher at Miller High. He felt confident succeeding among white people.
25. Sherry Parson (1958–1962), interview with the author, March 7, 2003.
26. See Finkelstein, "Is Adolescence Here to Stay? Historical Perspectives on Youth and Education"; and Fass, *The Damned and the Beautiful: American Youth in the 1920s*.
27. Almost all graduates across gender and race categories reported a social stigma attached to "dropping out" and parents' full expectation that their children would graduate from high school. All graduates—white and black, female and male—commented on the importance their parents placed on their high school education. For white and black males in the general track in particular, who were "doing time," it was their family members' insistence on a high school education and their own loyalties to their family that kept them in school.
28. At the dawn of the twentieth century, G. Stanley Hall, president of Clark University, and Charles W. Elliot, president of Harvard University, debated about the direction American high schools should take. The view that all students' needs should be addressed through a diversified curriculum became the template for the comprehensive high school. See G. Stanley Hall, "How Far Is the Present High School and Early College Training Adapted to the Nature and Needs of Adolescents?" *School Review* 9 (1901):

649–681. The report *Cardinal Principles of Education*, published by the Bureau of Education, U.S. Department of the Interior (Washington, D.C.: Government Printing Office, 1918), fully espoused Hall's argument on behalf of a diversified curriculum to keep as many students involved in high school life as possible.
29. James Coleman, *The Adolescent Society: The Social Life of the Teenager and Its Impact on Education* (NY: Free Press of Glencoe, 1961).
30. James Coleman, *The Adolescent Society: The Social Life of the Teenager and Its Impact on Education*, 217.
31. See Finkelstein, "Is Adolescence Here to Stay: Historical Perspectives on Youth and Education;" see also Robert Lynd and Helen Lynd, *Middletown* (NY: Harcourt Brace, 1929).
32. Status titles such as homecoming queen and May queen, as well as many of the extracurricular activities, were not student-generated in the 1950s and 1960s, but were handed down to students as integral parts of the school's modus operandi, further underscoring the weight of school structure in the lives of the Divided Generation. These were institutionalized performances.
33. Linda Moss (1965–1969), interview with the author, September 24, 2003.

3 Memories of Interracial Peer-Group Affiliations: Integration Years(1970–1985)

1. Joanne Pet (1970–1974), interview with the author, October 16, 2003.
2. Tim Whittle (1977–1981), interview with the author, June 28, 2004.
3. Tim Whittle (1977–1981), interview with the author, June 28, 2004. Also, Jeremy Garnes (1972–1976), interview with the author, September 9, 2003: "African Americans at the time were all local. They were local Miller Town people that we knew from growing up with." The term "local" is often used by Miller Town residents who have long roots in the community during casual conversations among neighbors.
4. See Table 3 in Methodology: The Transparent Historian.
5. Michelle Shaw (who began teaching at Miller High in 1960), interview with the author, November 7, 2003.
6. African American Teresa Randle (1977–1981), interview with the author, June 9, 2004.
7. Nat Right is the one graduate of the Divided Generation who referred to athletes of his time as jocks. The jock and the cheerleader can be identified as early as the 1920s, when football became an institutional feature of secondary education. See works by Paula Fass, *The Damned and the Beautiful: American Youth in the 1920s* (NY: Oxford University Press, 1977), and Elliott West, *Growing Up in Twentieth-Century America: A History and Reference Guide* (Westport, CT: Greenwood Press, 1996). The absence of football at Miller High until end of the 1960s is a cultural anomaly.

8. Jeremy Garnes (1974–1978), interview with the author, September 9, 2003: "I was a jock. That was the big thing. That's what most of us cared about." Also, David Randle (1972–1976), interview with the author, November 7, 2003: "We were jocks...it's [playing football] something I really wanted to do."
9. Tim Whittle (1977–1981), interview with the author, June 6, 2004. Also, David Randle (1972–1976), November 7, 2003: "I knew everybody!" Also, Jeremy Garnes, (1974–1978), September 9, 2003: "My first day was actually not at school. It was on the field."
10. Josh White (1972–1976), interview with the author, October 1, 2003.
11. Tim Whittle (1977–1981), June 6, 2004.
12. Ibid.
13. David Randle (1972–1976), November 7, 2003.
14. Ibid.
15. Ibid.
16. Ibid.
17. Ibid.
18. Most graduates interviewed, whether black or white, male or female, identified the fully integrated experience of the black jock at Miller High. The musician Michael Hallner (1972–1976), interview with the author, September 2003, explained: "If the black students were jocks, they stayed with [white] jocks."
19. David Randle (1972–1976), November 7, 2003.
20. Ibid.
21. Ibid.
22. Ibid.
23. Tim Whittle (1977–1981), interview with the author, June 28, 2004. Also, Jeremy Garnes (1974–1978), interview with the author, September 9, 2003: "My education was probably not the best at Franklin High. I knew it was too easy...today [my kids'] workload is very, very hard....I don't believe I had any homework that I remember...When I went to college it was a rude awakening."
24. Tim Whittle (1977–1981), interview with the author, June 28, 2004: "The rules were breaking down...We had release time, which means if you had enough credits, you didn't have to take certain amount of classes. I didn't take seventh period at all senior year. And one whole school day, I just had study period."
25. Historians have tracked this trend among white male students since the early decades of the twentieth century. See Paula Fass, *The Damned and the Beautiful: American Youth in the 1920s*.
26. Gerald Grant, in *The World We Created at Hamilton High*, alludes to teachers' laissez-faire practices during the 1970s at Hamilton High, a city high school. However, he attributes those lax practices to younger teachers' "guilty liberalism," teachers' general confusion about standards and discipline regarding black students under pressures of advocacy groups, and redefinitions of relationships between minors and adults with Supreme Court decisions such as the *Gault* case in 1967 and the *Winship* case in 1970. Of interest here is that

academic leniency was practiced at Miller High not by younger, but by older teachers who could not wait to retire.
27. Jeremy Garnes (1974–1978), September 9, 2003. Echoing Jim Garnes'experience, Tim Whittle (1977–1981) recalled: "It was two sets of teachers. You got your young ones and your older ones…You could go right down my grades and see who my teacher was…the older teachers were putting in their time."
"The older ones were the easy grade?"
"The older ones because they didn't give a hoot anymore. The younger ones challenged you."
28. David F. Labaree, *How to Succeed in School without Really Learning* (New Haven: Yale University Press, 1997).
29. Josh White (1972–1976), interview with the author, October 1, 2003.
30. Jeremy Garnes (1974–1978), interview with the author, September 9, 2003.
31. David Randle (1972–1976), interview with the author, November 11, 2003: "He [the coach] saw something in me and he helped me out. He really helped me a lot." Also, Tim Whittle (1977–1981), interview with the author, June 6, 2004: "He [the coach] was one of the nicest, best teachers I ever had. He and I really got along…my graduation, he made it a point to find my father and told my dad that I was one of the best students he had. That meant a lot to me…He was always in my corner." Also, Josh White (1972–1976), interview with the author, October 1, 2003: "Everybody liked him [the coach]."
32. Josh White (1972–1976), interview with the author, October 1, 2003.
33. African American alumna Teresa Randle (1977–1981), interview with the author, June 9, 2004.
34. Teresa Randle (1977–1981), interview with the author, June 9, 2004.
35. Sophie Baker (1981–1985), interview with the author, June 17, 2004.
36. Teresa Randle (1977–1981), interview with the author, June 9, 2004.
37. Sophie Baker (1981–1985), interview with the author, June 17, 2004.
38. Teresa Randle (1977–1981), interview with the author, June 9, 2004:
"I keep in touch with my teammates to this day!"
"A lot of bonding happens in sports?"
"It does!"
39. Of all the graduates interviewed during this period, Sophie, a 1985 graduate, was the only one to speak about the tangible markers of her life as a high school jock. Needless to say, only many more interviews might reveal whether this is a gender-specific recollection more akin to white female students, or if students of the 1970s were generally less interested in material possessions than Sophie, who graduated in the mid-1980s, might have been, reflecting a more materially conscious time. Then, too, Sophie's recollection might simply reflect her particular personality. Nevertheless, such details capture the known trend among U.S. high school students since the 1930s/1940s bobbysoxers of mass consumption of high-school-specific goods, from class rings to pens to pins. See Grace Palladino, *Teenagers: An American History*.
40. Teresa Randle (1977–1981), interview with the author, June 9, 2004: "If I had studied, I could have done better. I was just content with getting

by...Bs and Cs." Also, Sophie Baker (1981–1985), interview with the author, June 17, 2004: "I would actually calculate how much homework I could miss and still get a B...there were times when I didn't feel like doing it. I played sports."

41. Teresa Randle (1977–1981), interview with the author, June 9, 2004: "I didn't take advantage of the opportunity that was there. I did enough to get by." Also, Sophie Baker (1981–1985), interview with the author, June 17, 2004: "In retrospect...it was easy. Because it was easy, I didn't do everything I should have done."
42. Teresa Randle (1977–1981), interview with the author, June 9, 2004: "They didn't tell us about other colleges."
43. Sophie Baker (1981–1985), interview with the author, June 17, 2004.
44. Josh White (1972–1976), interview with the author, October 1, 2003: "When I was in school, you were either a head or a jock."
45. Ibid.
46. Jeremy Garnes (1974–1978), interview with the author, September 9, 2003.
47. Josh White (1972–1976), interview with the author, October 1, 2003.
48. Ibid. Josh White's use of the term "funny" was not expressed in a derogatory manner as if to laugh at the mailman, but in a manner that suggested surprise over the fact that the mailman is a peer who attended high school when he did. Nevertheless, the particular recollection suggested a "wasted" life.
49. Tim Whittle (1977–1981), interview with the author, June 28, 2004: "a buddy of mine, a football player, and he loved to get high. The teacher said, 'I'm going to make sure you don't graduate, you're a loser.'"
50. Sam Garnes (1970–1974), September 16, 2003.
51. A stereotype reinforced in popular movies such as *Grease*, where the car is used for "making out," engaging in illegal drag racing, or getting high.
52. From an informal conversation with Sam Garnes (1970–1974). Sam shared this information during our informal conversation that followed the tape-recorded conversation. Within about thirty-five minutes of recording, Sam asked that we turn off the recorder and continue our conversation without being recorded.
53. Joanne Pet (1970–1974), interview with the author, October 16, 2003
54. Carnoy and Levin quoted in David F. Labaree, *How to Succeed in School without Really Learning* (New Haven: Yale University Press, 1997), 49.
55. Pat Baley (1971–1975), interview with the author, October 20, 2003: "I was a musician...I had my own band, classic rock." Also, Michael Hallner (1972–1976), interview with the author, September 2003: "We [Michael and two other friends] got together, practiced, but never got it rollin' 'cause we all had a difference of opinion of music. They wanted to play Kiss...I wanted to play top forty, where you can make some money."
56. Tim Whittle (1977–1981), interview with the author, June 28, 2004.
57. Linda Moss (1965–1969), interview with the author, September 9, 2003.

4 Bridging Social Divides Through Peer-Groups: A Socially Tolerant but Politically Inactive Student Body (1970–1985)

1. Speaking of cross-racial relationships: African American alumnus Pat Baley ((1971–1975), interview with the author, October 20, 2003) said: "We got along. Most of the students you went to high school with, you started with in elementary school." Also, white alumnus Josh White, (1972–1976), interview with the author, October 1, 2003: "We got along really well." African American alumnus David Randle, (1972–1976), interview with the author, November 7, 2003: "My best friends were white." African American alumnae Joanne Pet and Teresa Randle and white alumna Sophie Baker reported as much.
2. Refer to African American alumna Teresa Randle's testimony in chapter 3 regarding her white friend and her KKK father, as well as to stories by African American alumnus David Randle and white alumnus Tim Whittle regarding white peers policing other white peers against racist attitudes.
3. See the work of Philip Cusick, *Inside High School: The Student's World* (NY: Holt, Rinehart, and Winston, 1973). Cusick argues that the small friendship groups of youth are reactions to the disempowering effects of secondary schools, which herd youth and devoid them of autonomy.
4. See Stephanie Coontz, *The Way We Never Were: American Families and the Nostalgia Trap* (NY: Basic Books, 2000), xi.
5. Tim Whittle (1977–1981), interview with the author, June 28, 2004.
6. Speaking of their principal, white alumnus Josh White (1972–1976), interview with the author, October 1, 2003: "I don't remember him when I was there." Also, African American alumna Teresa Randle (1977–1981), interview with the author, June 9, 2004: "Mr. Krauss was distant. I mean, he didn't really know me personally. He was distant."
7. The football jock culture might have participated in shifting aggression from male students against coaches (the reader will recall Robert Heart's testimony of his friend punching the coach and dropping out) to aggression by coaches against male students. For accounts of connections between sports and racial integration, see Pamela Grundy's acclaimed work, *Learning to Win: Sports, Education, and Social Change in Twentieth-Century North Carolina* (Chapel Hill: University of North Carolina, 2001).
8. A series of laws had been passed that expanded students' rights in secondary schools as well as the rights of young people in juvenile court proceedings. See rulings in *Tinker vs. DeMoines*, 1968; also in 1968, the ACLU published "Academic Freedom in the Secondary Schools," in which ACLU lawyers argue for "a recognition that deviation from the opinions and standards deemed desirable by the faculty is not ipso facto a danger to the educational

process"; see also Supreme Court decisions in the *Gault* case in 1967 and the *Winship* case in 1970. For details about *Tinker vs. DeMoines*, see Urban and Wagoner, *American Education: A History* (NY: McGraw Hill, 2000), 338. For details about the ACLU on academic freedoms in secondary education, and the *Gault* and *Winship* cases, see Gerald Grant, *Hamilton High*, 50–51.
9. Speaking about younger teachers: Tim Whittle (1977–1981) (interview with the author, June 28, 2004): "The younger ones challenged you. They were the ones you got into arguments with. They also stimulated you...If they had to put up with you being a bit playful or something, as long as it did not get out of hand, sometimes they would join in, they would find it funny."
10. Teresa Randle (1977–1981) (interview with the author, June 9, 2004): "I always had a good rapport with teachers." Also, Sophie Baker (1981–1985) (interview with the author, June 17, 2004): "I think a lot of people formed fairly close bonds with teachers at that time."
11. Sophie Baker (1981–1985), interview with the author, June 17, 2004.
12. Refer in part I, chapter 1, to Budd Land's remark (1964–68): "The general track was for those who were just going to go out and work as laborers. They were the slower kids or those that just didn't care about what they were doing."
13. Tim Whittle (1977–1981) reported other incidents of teachers calling students losers (see chapter 3). Pothead Sam Garnes (1974) and musician Michael Hallner (1986) also mentioned teachers' divisions of students into winners and losers.
14. For a succinct analysis of the social-efficiency perspective on education, see Chapter 1 in Labaree, *How to Succeed in School without Really Learning* (New Haven, CT: Yale University Press, 1997), 15–52.
15. Refer to Philip Cusick, *Inside High School: The Student's World*, earlier cited.
16. See Arthur G. Powell and Eleanor Farrar, *The Shopping Mall High School* (Boston: Houghton Mifflin, 1985), as well as Labaree's work earlier cited.
17. Kenneth Fish, *Conflict and Dissent in the High School*, 6.
18. E.F. Frazier, *Negro Youth at the Crossroads: Their Personality Development in the Middle States*, 105.
19. For works that examine black high school youth political engagement, see Dwayne C. Wright, "Black Pride Day, 1968: High School Student Activism in York, Pennsylvania," *Journal of African American History* 88 (2003): 151–162; Dionne Danns, "Chicago High School Students' Movement for Quality Education, 1966–1971," *Journal of African American History* 88 (2003): 138–150. See also V.P. Franklin, "Black High School Student Activism in the 1960s: An Urban Phenomenon?" *Journal of Research in Education* 10 (Fall 2000): 3–8.
20. Joanne Pet (1970–1974), interview with the author, October 16, 2003.

5 Memories of Segregation by Class, Race, Nationality, and Religion: Destabilizing Years of Shifting Demographics (1986–2000)

1. Heather Korn (1989–1993), interview with the author, February 25, 2003.
2. Sue Cohen (1991–1995), interview with the author, September 2003.
3. <Miller High> Alumni Directory (2000), xi.
4. At that time, many Asian Indians and Mexicans were also immigrating to Miller Town, but in much smaller numbers than Russian Jews (see Table 3). Because of limited resources available to conduct this oral history research and the larger number of Russian immigrants, I chose to concentrate my efforts on locating Russian alumni. See Addendum: Methodology.
5. <Miller Town> Community Times, throughout the 1970s. Holdings in archives of <Miller Town> library.
6. <Miller Town> Community Times, throughout the 1970s. Holdings in archives of <Miller Town> Library. Also, alumni recalled the transformations they had witnessed: Harry Rice (1983–1987), interview with the author, February 25, 2003: "It was strange to see so many new houses and roads that weren't there before." Cherry Gate (1988–1992), interview with the author, October 23, 2003: "My father, he's seen a complete change in the community since he's been there, and he's been there for thirty-one years. Yeah. And they built up around us…I mean, it used to be like all farmland…I forget about the malls…'cause Springfield Mall was just opening and Oak Field at the time."
7. Vera Debin (1995–1999), interview with the author, January 15, 2004.
8. Also, African American alumna Teresa Randle, 1981 graduate, compared the 1990s black student population to that of the 1970s: "It has changed now. I went to a graduation last week, there are more African American kids, almost 60 percent. I read in the paper that they have more racial tensions. I didn't witness that when I was in school [in the 1970s]."
9. David Randle (1972–1976), interview with the author, November 7, 2003. These reflections on changes in the community were shared by those alumni who were themselves "old-timers," and most had graduated from Miller High in the 1970s and early 1980s. David Randle, who is still a resident of Miller Town, has witnessed changes in the community firsthand.
10. Tim Whittle (1977–1981), interview with the author, June 28, 2004. Again, Tim Whittle counts among the "old-timers." There is a parallel here to the reaction of northern city blacks who felt that their good standing among the white community was endangered by the arrival of southern rural blacks. See Nicholas Lemann, *The Promised Land: The Great Black Migration and How It Changed America* (NY: Alfred A. Knopf, 1991).
11. African American alumna Teresa Randle, 1981 graduate, explained: "I have one child, she goes to a private school. So long as I can do it, I will keep her

there. I feel it is safer... There are a lot of kids that are angrier than ever before. Other people are raising them... We have kids in group homes in this area now... It's like foster homes, which means that more and more kids are on medication. When they don't take their medication you don't know what kind of child you will have that day. I just don't want my daughter to be involved in that." African American alumnus Pat Baley explained: "We homeschool. My kids are good and I want to keep them that way. Kids get all their bad habits starting in middle school. [I want] to protect [my kids] against bad influence of peers." White alumna Roberta Jones, 1991 graduate, explained her parents' choice regarding the schooling of her brother: "My brother went to a private school... [My parents] were worried the years that came under me, each year seemed to be a little bit worse and worse, worrying about whom your kids were hanging out with... He was a boy and he could get into more trouble."

12. Harry Rice (1983–1987), interview with the author, February 25, 2003. While the traditional ringing of the bell on the first day of school continued, and while the building remained, in the eyes of the Redivided Generation, an antediluvian construction, students' means and styles of communication were high tech. Administrators incorporated into school policies rules about the use of things such as cell phones. In the decade of Columbine-like tragedies, because of the insistence by many parents that their children be allowed to use cell phones on school premises, both as emergency tools as well as means for coordinating last-minute changes in overbooked schedules, and because of the speed at which drug deals could be arranged between students during school hours, and friends or lovers could chat away instruction time across classrooms through text messages, the cell phone emerged as "a tool so great," to quote the humorous Firesign Theater, "it could only be used for good or evil." Conference calls and text messaging further extended young people's ability to connect with each other in virtual communities. In the three-dimensional, physical world, Miller High School students now met at the recently opened mall in a neighboring town.

13. Heather Korn (1989–1993), interview with the author, February 25, 2003: "[There was] definitely kissing in the hallways... When you've seen it that often—we'd seen it since seventh and eight grade—they started way back. A girl, the summer of eighth grade, had twin boys... she was twelve actually. So we have been seeing it for a while."

14. Heather Korn (1989–1993), interview with the author, February 25, 2003: "If there was an alcohol or cigarette advertisement, they would be told to turn your shirt inside out, or they would have somebody come get you or bring you new clothes or something like that."

15. Ibid: "There was one kid, one or two classes ahead of us, who was very openly gay, and he got teased a lot... he got it a lot."

16. A member of FCA and of Young Life Club during his high school career, white alumnus Bill Jackman (1995–1999) (interview with the author, October 10, 2003) said, "We have a bible study." Also, Russian immigrant Vera Debin (1995–1999) (interview with the author, January 15, 2004): "Ultra right-wing [students attended] the morning bible study."

17. Of interest is that while gender no longer featured as a primary organizing category for 1990s alumni who consistently reported equal treatment of, interaction between, and participation by females and males, sexuality emerged as an organizing category. However, it was only briefly alluded to by one alumna.
18. All alumni interviewed had been "rule-abiding" students. While I was able to identify a student who had counted among those I have labeled "rule breakers," the alumnus refused to be interviewed or to suggest other alumni for me to interview. Thus "rule breakers" in this study come into view indirectly through recollections of those alumni interviewed, as well as through the recollections of two teachers whose oral histories I also collected and whose tenures reached back to the early 1960s through the early 1970s. See Table 1.
19. The following quotes illustrate the progressive decline of school spirit as we reach the end of the century. Harry Rice (1983–1987) (interview with the author, February 25, 2003): "[P]arents were involved...there was a lot of spirit." Also, Cherry Gate (1988–1992) (interview with the author, October 23, 2003): "We showed our school spirit...the band, and the choir, the auditorium was full. There was a lot of family, alumni who used to come and watch." Cecilia Hood (1992–1996) (interview with the author, October 7, 2003): "[School spirit] during homecoming week, but other than that, not really." Bill Jackman (1995–1999) (interview with the author, October 18, 2003): "I wouldn't say it was...school-spirited."
20. Sue Cohen (1991–1995), interview with the author, September 2003.
21. Bill Jackman (1995–1999), interview with the author, October 18, 2003.
22. Sue Cohen (1991–1995), interview with the author, September 2003.
23. As the Jewish student population grew at Miller High in the 1990s, more Jewish names appeared in the yearbook under pictures of the journalism club, suggesting a continuation into the end of the century, as the historian Paula Fass had identified, of proportionately more Jews, in particular female Jews, in extracurricular activities involving writing. See Paula in Joe Austin and Michael Willard, earlier cited.
24. Bill Jackman (1995–1999), interview with the author, October 18, 2003.
25. Bill Jackman (1995–1999) (interview with the author, October 18, 2003): "I remember one of my honors history teachers that was very demanding. We had to write a report on two presidents a week—four, five pages each president."
26. Sue's testimony underscores once again the weight of economic status in academic competition, given that attending college summer courses requires financial wherewithal.
27. Scans of yearbook captions next to senior pictures across time periods consistently reveal proportionately fewer black students in honors and gifted classes, which were referred to by earlier generations as "academic" tracks.
28. Although I was not able to secure interviews with several black students who graduated in the 1990s (see Addendum: Methodology) and whom I had identified through snowballing, I know firsthand through my personal friendship with many long-time black residents of Miller Town that many of the young black people in the community were and continue to

be actively involved in their churches. Russian immigrant alumna Vera Debin, 2000 graduate (interview with the author, January 15, 2004), recalled "black kids speaking of Jesus" in the hallways and talking about church meetings.
29. Ivan Strasky (1998–2002), interview with the author, November 6, 2004.
30. The guidance counselor at Miller High spoke highly of Ivan and recommended that I interview him to capture the Russian perspective.
31. Vera Debin (1995–1999) (interview with the author, January 15, 2004): "There was a very few teachers that I had any kind of relationship with other than in class." Cecilia Hood (1992–1996) (interview with the author, October 7, 2003): "I don't really remember the teachers that much."
32. Harry Rice (1983–1987) (interview with the author, February 25, 2003): "I was one of the geeks, no doubt."
33. Bill Jackman (1995–1999) (interview with the author, October 18, 2003): "The Russian kids were definitely a group." Cecilia Hood (1992–1996) (interview with the author, October 7, 2003): "People kind of stayed in groups…we had a large population of Russian kids." Betty Ames (1995–1999) (interview with the author, August 19, 2003): "Everyone definitely segregated themselves…two tables of…and then the Russian kids."
34. Ivan Strasky (1998–2002) (interview with the author, November 6, 2004): "Some [Russian] kids were athletes…I do remember two Russian kids who did wrestling."
35. Vera Debin (1995–1999), interview with author, January 15, 2004.
36. Ivan Strasky (1998–2002) (interview with author, November 6, 2004): "When Russian kids chose not to take the path of defensiveness, they [could] easily succeed." Vera Debin (1995–1999) (interview with author, January 15, 2004): "The programs were not difficult to get into; anyone who put in the effort would be able to do so. The people that were in standard, they chose to be there. They didn't want to put the extra effort. They were satisfied with their place."
37. Ivan Strasky (1998–2002), interview with the author, November 6, 2003.
38. Vera Debin (1995–1999), interview with the author, January 15, 2004.
39. Cherry Gate (1988–1992), interview with the author, October 23, 2003.
40. Heather Korn (1989–1993), interview with the author, February 25, 2003. Mr. L. presided over Miller High between 1978 and 1991, and his female successor, the only female principal in the history of Miller High during the twentieth century, presided between 1991 and 1997.
41. Bill Jackman (1995–1999), interview with the author, October 18, 2003.
42. Cherry Gate (1988–1992), interview with the author, October 23, 2003.
43. Bill Jackman (1995–1999) (interview with the author, October 18, 2003): "There were people in school that were racist either way."
44. Cecilia Hood (1992–1996), interview with the author, October 7, 2003.
45. Cherry Gate (1988–1992), interview with the author, October 23, 2003.
46. Heather Korn (1989–1993), interview with the author, February 25, 2003.
47. Heather Korn (1989–1993) (interview with the author, February 25, 2003): "All my friends were in standard."

48. Heather Korn (1989–1993) (interview with the author, February 25, 2003) echoed Cherry Gate's assessment of preps: "A lot of the kids that did hang out with the kids that did have the money, they would have the snobby attitude."
49. Heather Korn (1989–1993), interview with the author, February 25, 2003. Although Heather did not directly identify the group, and although it is impossible to extract a quote from the transcript which would confirm it, there is good reason to believe that she spoke of the Russian students, since she had identified all other groups by their names directly (blacks, preps, etc.), but had not mentioned the Russians. All other graduates of the Redivided Generation directly identified the Russians.
50. Cherry Gate (1988–1992) (interview with the author, October 23, 2003): "I worked for Riverside Veterinary Hospital my senior year...well, I used to groom when I was at horse shows, and I worked at my parents' farm." Roberta Jones (1987–1991) (interview with the author, September 2, 2003): "I worked as a grocery clerk on weekends and a couple of times during the week."
51. Heather Korn (1989–1993) (interview with the author February 25, 2003): "I cried the entire time...I'm pretty sentimental...I had a friend in every row...some of them knew they were going in the service...I had one friend whose dream it was to be a housewife." Cherry Gate (1988–1992) (interview with the author October 23, 2003): "Graduate. Move on with my life."
52. Roberta Jones (1987–1991), interview with the author, September 2, 2003.
53. Cherry Gate (1988–1992) (interview with the author, October 23, 2003): "A lot of the guys I hung out with were EMT and junior firefighters." Cherry Gate also recalled that her then-boyfriend, who is now her spouse, had been in the work-release program, and that he trained in a local mechanic shop. Heather Korn (1989–1993) (interview with the author, February 25, 2003): "The junior firefighters. You heard the siren go off, they all sat up straight, itching to get out of their seats."
54. Cecilia Hood (1992–1996), interview with the author, October 7, 2003.
55. Betty Ames (1995–1999), interview with the author, August 19, 2003. Also, Cecilia Hood (1992–1996) (interview with the author, October 7, 2003): "The popular kids, some of us played sports...we were friends since middle school."
56. Also Vera Debin (1995–1999) (interview with the author, January 15, 2004): "It was big limousines and big fancy dresses."
57. See Finkelstein earlier cited.
58. Cecilia Hood (1992–1996), interview with the author, October 7, 2003.
59. Sue Cohen (1991–1995), interview with the author, September 2003.
60. Vera Debin (1995–1999, interview with the author, January 15, 2004.
61. Betty Ames (1995–1999) (interview with the author, August 19, 2003): "I would sit in front of the TV from eight to eleven every night doing my homework. I'd do it in front of the TV. I didn't make straight As, I got by... When I went to college, I went *wow* [speaker's emphasis]...freshman year I did more work than I ever did in all four years of high school." Also Cecilia Hood

(1992–1996) (interview with the author, October 7, 2003): "I don't remember the teachers that much. I remember the social and the sports."
62. Refer to Addendum, Methodology: The Transparent Historian.
63. Vera Debin (1995–1999), interview with the author, January 15, 2004. I asked Vera, "Would black kids attend those [bible study] meetings?" Vera categorically answered, "No." I asked, "So this was just—," at which point Vera interrupted to say, "The ultra-religious right."
64. Sue Cohen (1991–1995) (interview with the author, September 2003): "Our football team wasn't so good." Also, Cecilia Hood (1992–1996) (interview with the author, October 7, 2003): "Football was horrible back then." Also Bill Jackman (1995–1999), interview with the author, October 18, 2003: "The football team didn't get a whole lot of support because they were one and nine every year while I was in high school."
65. Betty Ames (1995–1999) (interview with the author, August 19, 2003): "The soccer players were white, but there'd be an Indian kid who'd play soccer." This was corroborated by yearbook data. Further research should investigate the reasons for lack of black male students' participation in lacrosse and soccer.
66. For a discussion of the manifestations of punk subculture, from suburban youth whose encounters with punk are "primarily through videos on MTV and VH1" to the "genuine punks...who listen solely to independently produced anarcho-punk, steal or panhandle their means of subsistence...and eschew most forms of commodity," see Stacy Thompson, "Punk Cinema," *Cinema Journal* 43 (2004): 47.
67. Ibid.
68. For a discussion of the German roots of the goth subculture that emerged in the 1990s, see Gabriele Eckart, "The German Gothic Subculture," *German Studies Review* 28 (2005): 547–562. In this work, Eckart discusses a continuum among goths, who, while generally nonpolitical, have exhibited tendencies toward fascism at one extreme and antimilitarism at the other.
69. For a brief review of Afro-punk, see Black Grooves, "Afro-Punk," Archives of American Music and Culture, http://blackgrooves.org/?p=283 (accessed July 17, 2009).
70. Ivan Strasky (1998–2002), interview with the author, November 11, 2004.
71. David Randle (1976) alluded to African American pride when he stated that the new generation was no longer interested in cooperating, but rather in seeing faster change in the white status quo. Russian pride was alluded to by Ivan Strasky (2002): "They assert that image, the tough-guy image."
72. Ivan Strasky (1998–2002), interview with the author, November 6, 2004.
73. This sentiment was also echoed by an African American colleague who had graduated from high school in the early 1990s, and who said that African Americans no longer considered it acceptable that white researchers interview African Americans. My colleague's position is fully justified given the abusive ways that some white researchers have indeed used "research" to advance their theories of racial superiority, and the biased ways they have developed myriad explanations for minority youth's lower school achievements that bypass or

deny structural forces and institutionalized racism. Her remarks emerged from our discussions around the epistemology of ethnography and more vividly sensitized me to questions of legitimate knowledge-making. Because it is my deep commitment to contribute to scholarly work that helps those whose voices have been bypassed or silenced, and to help historical protagonists "appropriate the social construction of meaning to advance their own interests" (George W. Noblit et al., 2004, 14), I am called to be transparent. I am fully aware that to participate in exploring intergroup relationships through oral histories, I am to make transparent my middle-class whiteness so that whatever racial, class, or gender bias might infuse my work against my best efforts to remain as honest a researcher as possible, my colleagues—across racial, gender, class, or other divides—may more readily identify them as such. Oral historical research that explores multiple identities and border crossings among and between very diverse groups of people will always present the challenge of identity encounters between interviewer and interviewee. I am hopeful that as oral historians of education continue to make their researcher identities transparent, more honest articulations of our human investigations will emerge. For a thoughtful discussion of historical evolution of ethnographic inquiry in education and policy, and implications for emancipation of underrepresented voices, see George W. Noblit, Susana Y. Flores, and Enrique G. Murillon, Jr., *Postcritical Ethnography* (Cresskill, NJ: Hampton Press, 2004).
74. Ivan Strasky (1998–2002), interview with the author, November 6, 2004.
75. Ibid.
76. David Randle, a 1976 African American graduate, said that as a teacher and coach throughout the 1990s, he was more frequently breaking up fights between girls: "Now...fights every day, mostly girls."
77. Betty Ames (1995–1999), interview with the author, August 19, 2003.

6 Oppositional Self-Segregation: A Student Body Sensitized to Discrimination (1986–2000)

1. Vera Debin (1995–1999), interview with the author, January 15, 2004.
2. For a critical examination of the political power of elite parents to shape school curricula, and a discussion of how upper-class parents' ubiquitous presence in school classrooms, on school boards, and in myriad extracurricular activities, in the form of volunteer work and monetary contributions, buy them the political clout to lobby for ever more specialized top-end programs that privilege their children (GT, honors, AP classes), relegating the children of working-class and poor parents to the regular and remedial classes, see Peter Sacks, *Tearing Down the Gates: Confronting the Class Divide in American Education* (Berkeley and Los Angeles California: University of California Press, 2007).

3. Vera Debin (1995–1999), interview with the author, January 15, 2004.
4. Ivan Strasky (1998–2002), interview with the author, November 6, 2004.
5. Refer to testimonies of graduates of the Divided and Border-Crossing Generations in general. Also, in speaking of students who did not belong to peer groups during the 1970s and early 1980s: Joanne Pet (1970–1974) (interview with the author, October 16, 2003) recalled: "You segregated yourself regardless... It's just a natural thing. We do it every day. Subconsciously we do it. So, it's a comfortable type of segregation... Didn't you find that you could walk into the library and you had your Asian people kind of sitting here, black people sitting there... and you might have a Caucasian table here... You know, like on my job, we have maybe four Asian people, and they kind of do sit together in the canteen, they speak their own language... They feel more comfortable."
6. By no means do I suggest, as have historian Arthur Schlesinger and others, who believe that affirmations of difference ultimately further marginalize the already marginalized and "disunite America," that multicultural education in itself leads to fragmentation and division among students. However, findings in this history alert us to consider multicultural education through international perspectives, and further, to consider the role of broader institutional norms (such as ever-escalating tracking, uninvolved and distant faculty, etc.) in counteracting efforts to foster respect for difference through the curriculum.
7. Tyack and Cuban, *Tinkering Toward Utopia: A Century of Public School Reform* (Cambridge, MA: Harvard University Press, 1995), 44–45.
8. This finding was further corroborated by the testimony of Sam Buck, a Spanish teacher with more than twenty-five years of tenure who reflected on Miller High in the 1990s. Sam Buck, (interview with the author, November 7, 2003): "you had two parallel worlds"—the world of those whom Sam called the "mainstream" students, and those he placed in the category of "antisocial students." He said that students mostly went about their business—the mainstream not being bothered by the "troublemaker" constituency, and the troublemaker constituency not interested in the "mainstream" world.
9. Michelle Fine, Lois Weiss, and Linda C. Powell, *Harvard Educational Review* 67 (1997): 248.
10. See William Wraga, *Democracy's High School: The Comprehensive High School and Educational Reform in the United States* (Lanham, MD: University Press of America, 1994); see also Michael Sedlack, Christopher Wheeler, Diana C. Pullin, and Philip A. Cusick, *Selling Students Short: Classroom Bargains and Academic Reform in the American High School* (NY: Teachers' College Columbia University, 1986).

Conclusion

1. This finding is confirmed to some extent by research conducted by Amy Stuart Wells et al.: "We found more cross-racial friendships in high schools

with larger middle-class Black student populations." See "How Society Failed School Desegregation Policy: Looking Past the Schools to Understand Them," *Review of Research in Education*, 28 (2005): 83.
2. Findings by Amy Stuart Wells et al. also suggest "that there are several ways to interpret lunchroom segregation within desegregated schools...and one is to consider who students see and interact with outside of school." Wells (2005), 90. Certainly there are indications that even at the beginning of desegregation at Miller High, during the Divided Generation, interracial playmates of Miller Town were more likely to sit together in the cafeteria during lunchtime. Refer to Annie Cole's testimonies. I suggest that beyond imported habits of association, students also segregated in the cafeteria at the end of the century because of the absence of adult role-modeling and adult engagement.
3. Wells et al., "How Society Failed School Desegregation Policy: Looking Past the Schools to Understand Them," 90: "Close friendships within racially diverse schools are more often than not same-race friendships...." This historical analysis shows, however, that interracial friendships can flourish under particular social and demographic conditions such as those experienced by the Border-Crossing Generation.
4. Recent studies have documented patterns of school resegregation across the nation and warned of the dangers to a diverse democratic society of public practices that racially divide its youth. However, while these studies question the fate of the civil rights movements, they build arguments on the assumption that desegregated schools, by virtue of being desegregated, foster integrated environments. See report by Gary Orfield and Chungmei Lee, "Brown at Fifty: King's Dream or Plessy's Nightmare," Civil Rights Project Harvard University, http://www.gse.harvard.edu/news_events/features/2004/orfield01182004.html (accessed July 2, 2010).
5. Amy Stuart Wells, Jennifer Jellison Holme, Anita Tijerina Revilla, and Awo Korantemaa Atanda, "How Desegregation Changed Us: The Effects of Racially Mixed Schools on Students and Society," http://faculty.tc.columbia.edu/upload/asw86/ASWells041504.pdf (accessed July, 2009). On page five of the document, the authors state that they interviewed "more than 500 graduates, educators, advocates, and local policy makers who were directlyinvolved in racially mixed public high schools in different communities 25 years ago."
6. The state of Maryland, along with most other U.S. states in the 1990s, made changes to its laws governing juvenile justice and held more young offenders accountable through adult sentencing. See Megan C. Kurlycheck and Brian D. Johnson, "The Juvenile Penalty: A Comparison of Juvenile and Young Adult Sentencing Outcomes in Criminal Court," *Criminology* 42 (2004): 485–515. Wiley InterScience.
7. For examples of histories of education that address the high school's sorting mechanisms, see works by Samuel Bowles and Herbert Gintis, *Schooling in Capitalist America* (NY: Basic Books, 1976); James Anderson, *The Education of Blacks in the South, 1860–1935* (Chapel Hill, NC: University of North

Carolina Press, 1988); and William Reece, *The Origins of the American High School* (New Haven, CT: Yale University Press, 1995). According to Bowles and Gintis, school hierarchy of power and authority parallels the organization of power and authority in the workplace. The authors liken the role of grades to that of wages and establish a direct correspondence between competition among students and their lack of control over the curriculum with competition among workers and their lack of control over required contents of their assigned tasks. Bowles and Gintis advance that the American educational system serves the purposes and needs of the "production process and structure of class relations in the United States." In *The Education of Blacks in the South, 1860–1935*, Anderson identifies dominant paradigms of social reproduction in the Hampton Model of Normal School Industrial Education for blacks (Chapter 2) and tracks reproduction of castes in the black public high school of the South (Chapter 6). For reflections on the sorting mechanisms of public schools in general, see works by David Tyack, *The One Best System: A History of American Urban Education* (Cambridge, MA: Harvard University Press, 1974); David Tyack and Elizabeth Hansot, *Managers of Virtue: Public School Leadership in America, 1820–1980* (NY: Basic Books, 1982); and Michael Katz, *The Irony of Early School Reform* (NY: Teachers College Press, 1968). These historians of education identify the modernizing tendency of schools to prepare youth for various workstations in society, to assure that youth is "properly socialized to the new modes of production, attuned to hierarchy, affective neutrality, role-specific demands, extrinsic incentives for achievement" (Tyack, 1974, p. 73). For ethnographies that focus on the high school and expose its divisive and exclusionary practices, see Michelle Fine, *Framing Dropouts* (Albany: State University of New York Press, 1991); Ellen Bratlinger, *The Politics of Class in Secondary School* (NY: Teachers College Press, 1993); Jay MacLeod, *Ain't No Makin' It* (Boulder, CO: Westview Press, 1995; Anne Locke-Davidson, *Making and Molding Identity in School* (NY: SUNY Press, 1996). See also Jeanie Oakes, *Keeping Track: How Schools Structure Inequality* (New Haven, CT: Yale University Press, 2006); Jonathan Kozol, *Savage Inequalities: Children in America's Schools* (NY: Harper Collins, 1991). These ethnographies expose institutionalized racism and classism.

8. Anthropologist John Ogbu attributes student behavior within school to cultural influences outside of school. By comparing involuntary minorities' cultural backgrounds, Ogbu concludes that involuntary minorities (e.g., black youth) inherit through "family and community discussions and gossip, as well as through public debates over minority education" mistrust of school and school authorities and a deeply ingrained skepticism about their chances to succeed in schools. His work "Voluntary and Involuntary Minorities: A Cultural-Ecological Theory of School Performance with Some Implications for Education," *Anthropology and Education Quarterly* 29 (1998): 155–188, has received serious criticism. My own analysis challenges the notion that immigrants do not share with those whom Ogbu calls "involuntary minorities" the same distrust of school authorities, as

9. Schofield (1989), quoted in Amy Stuart et al., "How Society Failed School Desegregation Policy: Looking Past the Schools to Understand Them," *Review of Research in Education* 28 (2005): 71.
10. Jencks, quoted in Amy Stuart Wells et al., "How Society Failed School Desegregation Policy: Looking Past the Schools to Understand Them," *Review of Research in Education* 28 (2005): 71.
11. Gerald Grand, *The World We Created at Hamilton High*, 210.
12. Patrick Ryan, "A Case Study in the Cultural Origins of a Superpower: Liberal Individualism, American Nationalism, and the Rise of High School Life, A Study of Cleveland's Central and East Technical High Schools, 1890–1918," *History of Education Quarterly*, 45 (Spring 2005): 67.
13. Patrick Ryan, "A Case Study in the Cultural Origins of a Superpower: Liberalism, Individualism, American Nationalism, and the Rise of High School Life, A Study of Cleveland's Central and East Technical High Schools," *History of Education Quarterly*, 45 (Spring 2005): 94.

Addendum

1. See Donald Ritchie, *Doing Oral History* (New York: Twayne Publishers, 1995).
2. Alessandro Portelli states: "[T]he first thing that makes oral history different, is that it tells us less about *events* than about their *meaning*... what informants believe is indeed a historical *fact* (that is, the fact that they believe it), as much as what really happened." Alessandro Portelli, "What makes oral history different," in *The Oral History Reader*, Robert Perks and Alistair Thomson, eds. (New York: Routledge, 2005), 63–74. See also Alessandro Portelli, *The Death of Luigi Trastulli and Other Stories: Form and Meaning in Oral History* (Albany: SUNY Press, 1991).
3. See Penny Summerfield, *Reconstructing Women's Wartime Lives: Discourse and Subjectivity in Oral Histories of the Second World War* (Manchester: Manchester University Press, 1998); Alessandro Portelli, *The Death of Luigi Trastulli and Other Stories: Form and Meaning in Oral History* (Albany: SUNY Press, 1991).
4. See "Ethnicity and National Identity," special issue, *Oral History* 21 (1993); Susan K. Burton, "Issues in Cross-Cultural Interviewing: Japanese Women in England," *Oral History* 31 1(2003); Jelena Cvorovic, "Gypsy Oral History in Serbia: From Poverty to Culture," *Oral History* 33 1 (2005); "Black History," special issue, *Oral History* 8 1(1980).
5. Paul Thompson, *The Voice of the Past: Oral History* (Oxford: 1978), 7–8.

6. Ronald J. Grele, "Movement without Aim: Methodological and Theoretical Problems in Oral History," in *The Oral History Reader*, Robert Perks and Alistair Thomson, eds. (New York: Routledge, 2005), 38–53.
7. Mikhail Bakhtin, quoted in Donald Freeman, "To Take Them at Their Word: Language Data in the Study of Teachers' Knowledge," *Harvard Educational Review* 66 (1996): 748–750.
8. See Patricia Collins, *Black Feminist Thought: Knowledge, Counsciousness, and the Politics of Empowerment* (New York: Routledge, 2000); Patricia Collins, *Fighting Words: Black Women and the Search for Justice* (Minneapolis: University of Minnesota Press, 1998).
9. Leslie McCall, "The Complexity of Intersectionality," *Signs: Journal of Women in Culture and Society* 30 (2005): 1771.
10. Leslie McCall, "The Complexity of Intersectionality," *Signs: Journal of Women in Culture and Society* 30 (2005): 1771–1800.
11. "Intersectionality," http://en.wikipedia.org/wiki/Intersectionality (accessed June 15, 2009).
12. One of many examples of documentary data that contextualized oral testimonies involved the Russian middle-class alumna Vera Debin, a member of the Redivided Generation (1986–2000). Debin shared her feelings about the yearbook pictures—"they [the preps] went around and tried to get pictures of all the clubs, and all the band, but it was pretty much them and their friends, all extra pictures"—and further suggested that yearbooks would "probably be more interesting if people were assigned to a committee to do the yearbook, randomly selected" to ensure broad and fair representations of all students. I subsequently examined yearbooks for redundancy in pictures. Indeed, the same faces would reappear on the pages of the yearbooks. When graduates recalled what kind of student attended what kind of academic track, extracurricular activity, or sport, I systematically analyzed yearbooks for the distribution of students across extracurricular activities and curricular tracks, by race, ethnicity, and gender.
13. Irma Omeldo, "Redefining Culture through the Memorias of Elderly Latinas," *Qualitative Inquiry* 5 (1993): 353–376.
14. Constructing minimum or set numbers of testimonies becomes even more problematic when one considers the disparity in numbers between white and black students in the early stages of desegregation. For example, how many of the total 121 white students who graduated in 1958 should one interview to "counterbalance" interviews with the four African American students who graduated the same year? In relational histories of education, "difference" becomes a more relevant and workable criterion of selection than numbers and volume.
15. See works by Louis S. Diggs, *Since the Beginning: African American Communities in Towson* (Baltimore: Uptown Press, 2000); *Holding on to Their Heritage* (Catonsville, MD: privately printed, 1996); *In Our Own Voices: A Folk History in Legacy* (Catonsville, MD: privately printed, 1998).
16. I have since discovered the NviVo coding program, which I consider valuable for coding interviews.

17. See Paul Thompson, *The Voice of the Past: Oral History* (Oxford: Oxford University Press, 1988), 131.
18. Paula Fass, *Generations of Youth: Youth Culture and History in Twentieth-Century America*, 95–117.
19. Michael Frisch, *A Shared Authority: Essays on the Craft and Meaning of Oral and Public History* (New York, 1990), xx.

Bibliography

Anderson, James. *The Education of Blacks in the South, 1860–1935*. Chapel Hill, N.C.: University of North Carolina Press, 1988.
Anderson, James and Dara Byrne, eds. *The Unfinished Agenda of Brown v. Board of Education*. Hoboken, NJ: John Wiley & Sons, 2004.
Ashcroft, Bill, Gareth Griffiths, and Helen Tiffin, eds. *Post-Colonial Studies: The Key Concepts*. NY: Routledge, 2005
Austin, Joe, and Michael Willard, eds. *Generations of Youth: Youth Culture and History in Twentieth Century America*. NY: New York University Press, 1998.
Bailey, Beth. "From Panty Raids to Revolution: Youth and Authority, 1950–1970." In *Generations of Youth: Youth Culture and History in Twentieth Century America*, edited by Joe Austin and Michael Willard, 187–204. NY: New York University Press, 1998.
———. *From Front Porch to Back Seat: Courtship in Twentieth Century America* Baltimore: The Johns Hopkins University Press, 1998.
Black Grooves. Afro-Punk. Archives of American Music and Culture. http://blackgrooves.org/?p=283 (accessed July 17, 2009).
Boli, John, Francisco O. Ramirez, and John W. Meyer, "Explaining the Origins of Expansion of Mass Education." *Comparative Education Review* 29 (1985): 145–170.
Bowles, Samuel, and Herbert Gintis. *Schooling in Capitalist America*. NY: Basic Books, 1976.
Bratlinger, Ellen, A. *The Politics of Class in Secondary School*. NY: Teachers College Press, 1993.
Brown, Roger. *Social Psychology*. NY: Free Press, 1986.
Burke, Peter. *History and Social Theory*. NY: Cornell University Press, 2005.
Clayborne Carson, "Two Cheers for Brown v Board of Education," *Journal of American History* 91 (2004): 26–31.
Chism, Khalil. "A Documentary History of Brown." In *The Unfinished Agenda of Brown v. Board of Education,* edited by James Anderson and Dara Byrne, 7–22. Hoboken, New Jersey: John Wiley & Sons, 2004.
Cohen, Ronald. "The Delinquents: Censorship and Youth Culture in Recent U.S. History." *History of Education Quarterly* 37, no. 3 (1997): 251–270.
Collins, Patricia. *Black Feminist Thought: Knowledge, Consciousness, and the Politics of Empowerment*. NY: Routledge, 2000.

Collins, Patricia. *Fighting Words: Black Women and the Search for Justice.* Minneapolis: University of Minnesota Press, 1998.

Coleman, James. *The Adolescent Society: The Social Life of the Teenager and its Impact on Education.* NY: The Free Press of Glencoe, 1961.

Conant, James, B. *The American High School Today: A First Report to Interested Citizens.* NY: McGraw-Hill, 1959.

Coontz, Stephanie. *The Way We Never Were: American Families and the Nostalgia Trap.* NY: Basic Books, 2000.

Crenshaw, Kimberlé, Neil Gotanda, Gary Peller, and Kendall Thomas, eds. *Critical Race Theory: The Key Writings That Formed the Movement.* NY: New York Press, 1995.

Danns, Dionne. "Chicago High School Students' Movement for Quality Education, 1966–1971." *The Journal of African-American History* 88 (Spring 2003): 138–150.

Davidson, Locke Anne. *Making and Molding Identity in Schools.* NY: State University of New York Press, 1996.

DeHar, Hu. "An Asian American Perspective on Brown." In *The Unfinished Agenda of Brown v. Board of Education*, edited by James Anderson and Dara Byrne, 108–119. Hoboken, NJ: John Wiley & Sons, 2004.

Diggs, Louis. *Since the Beginning: African American Communities in Towson.* Baltimore: Uptown Press, 2000.

———. *In Our Own Voices: A Folk History in Legacy.* Catonsville, MD: Privately Printed, 1998.

———. *Holding On To Their Heritage.* Catonsville, MD: Privately Printed, 1996.

Dorn, Sherman. "Origins of the "Dropout Problem," *History of Education Quarterly* 33, no. 3 (1993): 353–373.

Eckart, Gabriele. "The German Gothic Subculture." *German Studies Review* 28 (2005): 547–562.

Eckert, Penelope. *Jocks & Burnouts: Social Categories and Identity in the High School.* NY: Teachers College Press,1989.

Eisenman, Linda. *Higher Education for Women in Postwar America, 1945–1965: Reclaiming the Incidental Student.* Baltimore: Johns Hopkins University Press, 2006.

Eisner, Elliot, W. and Alan Peshkin, eds. *Qualitative Inquiry in Education: The Continuing Debate.* NY: Teachers College, 1990.

Fass, Paula. "Creating New Identities: Youth and Ethnicity in New York City High Schools in the 1930s and 1940s." In *Generations of Youth: Youth Culture and History in Twentieth Century America,* edited by Joe Austin and Michael Willard, 95–117. NY: New York University Press, 1998.

———. *The Damned and the Beautiful: American Youth in the 1920s.* NY: Oxford University Press, 1977.

Fine, Michelle. *Framing Dropouts.* Albany: State University of New York Press, 1991.

Fine, Michelle, Lois Weiss, and Linda C. Powell. "Communities of Difference: A Critical Look at Desegregated Spaces Created by and for Youth". *Harvard Educational Review* 67 (Summer 1997): 247–285.

Finkelstein, Barbara. "Is Adolescence Here to Stay?: Historical Perspectives on Youth and Education." In *Adolescence and Society*, edited by Timothy C. Urdan and Frank Pajares, 1–33. NY: Information Age Press, 2003.

Fish, Kenneth. *Conflict and Dissent in the High School*. NY: Bruce Publishing Company, 1970.

Franklin, V.P. "Black High School Activism in the 1960s: An Urban Phenomenon?" *Journal of Research in Education*. 10, no.1 (2000): 3–8.

Frazier, E.F. *Negro Youth at the Crossroads: Their Personality Development in the Middle States*. NY: Scholar Books, 1967.

Freeman, Donald. "'To Take Them at Their World': Language Data in the Study of Teachers' Knowledge." *Harvard Educational Review* 66 (Winter 1996): 732–761.

Frisch, Michael. *A Shared Authority*. Albany: State University of New York Press, 1990.

Gilbert, John. *Cycles of Outrage: America's Reaction to the Juvenile Delinquent in the 1950s*. NY: Oxford University Press, 1986.

Gillis, John. *Youth and History: Tradition and Change in European Age Relations 1779-Present*. NY and London: Academic Press, Inc., 1974.

Grant, Gerald. *The World We Created at Hamilton High*. Cambridge: Massachusetts: Harvard University Press, 1988.

Graebner, William. *Coming of Age in Buffalo: Youth and Authority in the Postwar Era*. Philadelphia: Temple University Press, 1990.

Grele, Ronald, J. "Movement without Aim: Methodological and Theoretical Problems in Oral History." In *The Oral History Reader*, edited by Robert Perks and Alistair Thomson, 38–53. NY: Routledge, 2005.

Grundy, Pamela. *Learning to Win: Sports, Education, and Social Change in Twentieth Century North Carolina*. Chapel Hill: University of North Carolina, 2001.

Gutmann, Amy. *Democratic Education*. Princeton, NJ: Princeton University Press, 1999.

Hall, Stanley. "How Far Is the Present High School and Early College Training Adapted to the Nature and Needs of Adolescents?" *School Review* 9 (1901): 649–681.

Hollingstead, August, B. *Elmtown's Youth: The Impact of Social Class and Adolescents*. NY: John Wiley & Sons, 1949.

Irons, Peter. *Jim Crow's Children: The Broken Promise of the Brown Decision*. New York, NY: Viking, 2002.

Jackson, Kenneth, T. *Crabgrass Frontier: The Suburbanization of the United States*. Oxford, NY, Toronto: Oxford University Press, 1985.

Katz, Michael. *Class, Bureaucracy, and the Schools*. NY: Praeger, 1971.

———. *The Irony of Early School Reform*. NY: Teachers College, 1968.

Kett, Joseph. *Rites of Passage: Adolescence in America, 1790 to the Present*. New York: Basic Books, 1977.

Kozol, Jonathan. *Savage Inequalities: Children in America's Schools*. NY, Harper Collins, 1991.

Kurlycheck, Megan, C. and Brian D. Johnson. "The Juvenile Penalty: A Comparison of Juvenile and Young Adult Sentencing Outcomes in Criminal

Court." *Criminology* 42 (2006): 485–515, http://www3.interscience.wiley.com/journal/118749263/abstract?CRETRY=1&SRETRY=0

Labaree, David. *The Making of an American High School: The Credentials Market and the Central High School of Philadelphia, 1838–1939.* New Haven: Yale University Press, 1988.

———. *How to Succeed in School Without Really Learning.* New Haven: Yale University Press, 1997.

Lemann, Nicholas. *The Promised Land: The Great Black Migration and How It Changed America.* NY: Alfred A. Knopf, Inc., 1991.

Levine, Robert and Donald Campbell. *Ethnocentrism: Theories of Conflict, Ethnic Attitudes, and Group Behavior.* NY: John Wiley & Sons, 1972.

Levesque, Roger, JR. *Dangerous Adolescents, Model Adolescents: Shaping the Role and Promise of Education.* NY: Kluwer Academic/Plenum Publishers, 2002.

Lynd, Robert and Helen Lynd. *Middletown.* NY: Harcourt Brace, 1929.

MacDonald, Victoria-María, ed. *Latino Education in the U.S.: A Narrative History, 1531–2000.* NY: Palgrave/Macmillan, 2004.

MacLeod, Jay. *Ain't no Makin' It.* Boulder, CO: Westview Press, 1995.

McCall, Leslie. "The Complexity of Intersectionality." Signs: Journal of Women in Culture and Society 30 (2005): 1771–1800, http://www.journals.uchicago.edu/doi/abs/10.1086/426800

Marginson, Simon. "After Globalization, Emerging Politics of Education." *Journal of Education Policy* 14 (1999): 19–35.

Modell, John. *Into One's Own: From Youth to Adulthood in the United States, 1920–1975.* Berkeley: University of California Press, 1989.

Noblit, George, Susana Y. Flores, and Enrique G. Murillon, Jr. *Postcritical Ethnography.* Cresskill, NJ: Hampton Press, Inc., 2004.

Oakes, Jeanie. *Keeping Track: How Schools Structure Inequality.* New Haven, CT: Yale University Press, 2006.

Olsen, Laurie. *Made in America: Immigrant Students in our Public Schools.* NY: The New Press, 1997.

Ogbu, John. "Voluntary and Involuntary Minorities: A Cultural-Ecological Theory of School Performance with Some Implications for Education." *Anthropology and Education Quarterly* 29 no. 2 (1998): 155–188.

Omeldo, Irma. "Redefining Culture through the Memorias of Elderly Latinas." *Qualitative Inquiry* 5 (1993): 353–376.

Orfield, Gary and Chungmei Lee. "Brown at Fifty: King's Dream or Plessy's Nightmare." The Civil Rights Project Harvard University. (Jan. 17, 2004), http://www.%20.civilrightsproject.harvard.edu/research/reseg40/resegregation04.php

Orfield, Gary. "Schools More Separate: Consequences of a Decade of Resegregation." Harvard University, The Civil Rights Project Web site, http://www.civilrightsproject.harvard.edu/research/deseg/separate_schools01.php

Palladino, Grace. *Teenagers: An American History.* Chicago: University of Chicago Press, 1996.

Palumbos, Robert. "Student Involvement in the Baltimore Civil Rights Movement, 1953–63." *Maryland Historical Magazine* 94 (1999): 449–492.

Peshkin, Alan. *Places of Memory: Whiteman's Schools and Native American Communities*. Mahwah, NJ: Lawrence Erlbaum Associates, 1997.

Portales, Marco. "A History of Latino Segregation Lawsuits." In *The Unfinished Agenda of Brown v. Board of Education*, edited by James Anderson and Dara Byrne, 123–136. Hoboken, NJ: John Wiley & Sons, Inc., 2004.

Portelli, Alessandro. "What Makes Oral History Different." In *The Oral History Reader*, edited by Robert Perks and Alistair Thomson, 63–74. NY: Routledge, 2005.

———. *The Death of Luigi Trastulli and Other Stories: Form and Meaning in Oral History*. Albany: SUNY Press, 1990.

Portes, Alejandro and Rubén G. Rumbaut. *Immigrant America: A Portrait*. Berkeley and Los Angeles: The University of California Press, 1996.

Powell, Arthur, Eleanor Farrar, and David Cohen. *The Shopping Mall High School*. Boston: Houghton Mcfflin Co., 1985.

Price, Jeremy. *Against the Odds: The Meanings of School and Relationships in the Lives of Six Young African-American Men*. Stamford, Connecticut: Alex Publishing Corporation, 2000.

Reese, William. *The Origins of the American High School*. New Haven: Yale University Press, 1995.

Ritchie, Donald. *Doing Oral History*. NY: Twayne Publishers, 1995.

Ryan, Patrick J. "A Case Study in the Cultural Origins of a Superpower: Liberal Individualism, American Nationalism, and the Rise of High School Life, A Study of Cleveland's Central and East Technical Schools, 1890–1918." *History of Education Quarterly* 45 (Spring 2005): 66–95.

Sacks, Peter. *Tearing Down the Gates: Confronting the Class Divide in American Education*. Berkeley and Los Angeles California: University of California Press, 2007.

Salomone, Rosemary, C. *Vision of Schooling: Conscience, Community and Common Education*. New Haven: Yale, 2000.

Sedlack, Michael, et al. *Selling Students Short: Classroom Bargains and Academic Reform in the American High School*. NY: Teachers' College Columbia University, 1986.

Shlossman, Stephen. *Love and the American Delinquent*. Chicago: University of Chicago Press, 1977.

Summerfield, Penny. *Reconstructing Women's Wartime Lives: Discourse and Subjectivity in Oral Histories of the Second World War*. Manchester: Manchester University Press, 1998.

Talburt, Susan. "Time, Space, and Ethnography Without Proper Subjects." In *Postcritical Ethnography*, edited by George Noblit, Susana Flores, and Enrique Murillo, 107–123. Cresskill, NJ: Hampton Press, Inc., 2004.

Thompson, Paul. *The Voice of the Past: Oral History*. Oxford: Oxford University Press, 1988.

Thompson, Stacey. "Punk Cinema." *Cinema Journal* 43 no.2 (2004): 47–66, http://muse.jhu.edu/login?uri=/journals/cinema_journal/v043/43.2thompson.html

Tyack, David and Larry Cuban. *Tinkering Toward Utopia: A Century of Public School Reform*. Cambridge, Massachusetts: Harvard University Press, 1995.

Tyack, David and Elizabeth Hansot. *Managers of Virtue: Public School Leadership in America, 1820–1980*. NY: Basic Books, 1982.

Urban, Wayne and Jennings Wagoner. *American Education*. NY: McGraw Hill Companies, Inc., 2000.

Vicino, Thomas, J. *Transforming Race and Class in Suburbia: Decline in Metropolitan Baltimore*. NY: Palgrave Macmillan, 2008.

Wells Stuart, Amy, et.al. "How Society Failed School Desegregation Policy: Looking Past the Schools to Understand Them." *Review of Research in Education* 28 (2005): 47–99.

———. "How Desegregation Changed Us: The Effects of Racially Mixed Schools on Students and Society." http://faculty.tc.columbia.edu/upload/asw86/ASWells041504.pdf

West, Elliot. *Growing Up In Twentieth Century America: A History and Reference Guide*. Westport, Connecticut: Greenwood Press, 1996.

Willis, Paul. *Learning to Labor*. NY: Columbia Press, 1977.

Wraga, William. *Democracy's High School: The Comprehensive High School and Educational Reform in the United States*. Lanham, Maryland: University Press of America, 1994.

Wright, Dwayne, C. "Black Pride Day, 1968: High School Student Activism in York, Pennsylvania." *The Journal of African-American History* 88 (Spring 2003): 151–162.

Zinn, Howard. *A People's History of the United States*. NY: HarperCollins Inc., 2003.

Index

academic achievement, 2, 33, 54, 152
administrators, 6, 17, 18, 23, 29, 48, 94, 95, 105, 136, 138, 139, 142, 202
adolescence, 8, 117, 178, 193, 194, 195
African Americans, xviii, 3, 4, 5, 11, 17, 18, 27, 30, 32, 33–65, 72–98, 122–144, 151, 153, 161, 162, 164, 168, 169, 171, 173, 174, 179, 180, 183, 190, 194, 195, 197, 199, 200, 201, 202, 206, 207, 212
see also blacks
assimilation, 12, 176, 177
associative living, xviii

Bakhtin, Mikhail, 158, 212
Baltimore city, 3, 172, 174
Baltimore County, xii, xvii, 1, 3, 17, 18, 30, 35, 165, 172, 190
Bilingual Education Act, xvi
blacks, xii, xiii, xv, xviii, 1, 3–6, 9–12, 13, 18, 19, 26, 27, 30, 31–68, 73, 90–110, 115, 116, 121–137, 138–140, 143–150, 151–155, 159, 161–164, 166, 170, 171, 173, 174, 175, 177, 179, 183, 190, 191, 192, 193, 194, 196, 200, 201, 203, 204, 205, 206, 209, 210, 211, 212
black, middle-class, 10, 31, 80, 150
black, women, 33, 34, 38, 39, 40, 47, 51–55, 88, 89, 97, 150, 151, 190
black feminist theory, 6, 159, 180, 212, 215

black men, 35, 36, 38, 40, 41, 46, 48, 49, 53, 65, 68, 82, 97, 98, 150, 173
see also African Americans
Brown V. Board of Education, xii, xv, 8, 9, 145, 175, 182

Cafeteria, 10, 11, 32, 52, 57, 64, 78, 87, 88, 89, 110, 111, 122, 130, 139, 142, 144, 148, 150, 151, 152, 182, 209
Carnoy, Martin and Levin, Henry, 9, 57, 142, 150, 182
Christian, 37, 49, 50, 105, 107, 109, 110, 113, 122, 126, 133, 134, 142, 145
city, xiii, xvi, xvii, 2–5, 8, 9, 10, 18, 40, 62, 98, 103, 104, 124, 128, 129, 130, 135, 136, 137, 139, 145, 148, 153, 172, 178, 185, 201
civil rights, xii, xv, xvii, 3, 4, 5, 8, 10, 87, 88, 147, 149, 154, 175, 176, 179, 182, 209, 218
class (economic status), xii, xiii, xv, xvi, xviii, 1, 2, 4–13, 17–90, 95–97, 99, 105, 108, 114, 117, 119, 121, 122, 126, 131, 133, 135, 137, 138, 139, 140, 145, 147, 148, 150, 151, 152, 153, 155, 157–161, 166, 167, 168, 170, 178, 179, 180, 183, 188, 193, 207, 209, 210, 217
lower-class, 10, 21, 24, 27, 28, 38, 43, 46, 56, 84, 85, 96, 97, 133, 139, 147, 166, 167, 168, 169, 207, *see also* working-class students

class (economic status)—*Continued*
  middle-class, 4, 10, 12, 17–57, 66, 71, 72, 78, 80, 82, 83, 84, 90, 107, 108, 111, 114, 117, 119, 122, 127, 133, 138, 139, 140, 147, 148, 150, 153, 166, 168, 169, 183, 193, 207, 209
classroom, xviii, 19, 20, 22, 28, 31, 36, 46, 50, 72, 77, 80, 94, 98, 114, 151, 187, 202, 207, 208
coding, 157, 170, 212
Coleman, James, 55, 175
Collins, Patricia, 6, 159, 177
community, 4, 5, 10, 11, 17, 19, 21, 24, 25, 28, 33, 36, 38, 48, 49, 52, 76, 96, 98, 103, 106, 107, 112, 117, 118, 119, 128, 129, 130, 139, 142, 147, 148, 150, 151, 153, 158, 160, 161–163, 165, 172, 177, 182, 184, 189, 191, 195, 201, 203, 210
comprehensive public school, xvi, 154
conservative students / town, xvii, 5, 10, 18, 45, 47–57, 126, 134, 138, 142, 143, 146, 147, 148, 181, 192
consumerism, 126, 131, 145
cross-group relationships, 1, 6, 7, 9, 10, 159, 160, 161, 171
  cross-class relationships, 2, 4, 5, 8, 10, 13, 52, 76, 138, 148
  cross-racial relationships, 52, 76, 109, 150, 199, 208
culture, xv, 9, 10, 13, 17, 23, 25, 28, 43, 72, 78, 97, 113, 114, 122, 129, 136, 148, 151–152, 153, 154, 158, 159, 160, 171, 178, 192, 195, 208, 210

dances, xviii, 19, 22, 38, 39, 41, 42, 49, 50, 85, 90, 108, 121, 151, 193
dating, 22, 23, 49, 50, 56, 91, 185
democracy, xv, xvi, xviii, 1, 20, 154, 155, 181
demographics, xviii, 4, 11, 18, 83, 103, 133, 134, 135, 148, 152, 153

desegregation, xii, xiii, xv, xvii, xviii, 1, 2, 3, 11, 12, 17, 18, 19, 30, 33, 34, 36, 37, 38, 39, 41, 42, 43, 48, 50, 51, 52, 56, 62, 74, 96, 107, 125, 138, 143, 144, 149, 150, 153, 158, 164, 167, 168, 175, 177, 178
diploma, 20, 23, 27, 33, 37, 55, 75, 81, 82, 118, 165, 185
discipline, 48, 70, 78, 94, 106, 108, 138, 196
discrimination, 41, 51, 76, 80, 88, 113, 125, 128, 142, 143, 144, 150, 151, 153

economic status, xv, xvi, 11, 18, 29, 42, 43, 44, 48, 52–55, 62, 70, 83, 84, 90, 99, 105, 113, 114, 117, 119, 121, 122, 126, 128, 138, 148, 151, 152, 154, 155, 159–163, 166, 176
Eisenman, Linda, 50, 183
Elementary and Secondary Education Act, xvi
ethnicity, xv, xvi, xviii, 1, 6, 12, 122, 133, 136, 137, 152, 154, 159, 161, 173
expulsions, 105, 106, 122, 141, 145
extracurricular activities, xvi, xviii, 10, 18, 21, 22, 40, 46, 47, 52, 67, 68, 80, 83, 98, 99, 108, 112, 119, 123, 129, 135, 147, 148, 173, 178, 192, 195, 203, 207

family, 19, 21, 22, 23, 34, 36, 43, 46, 48, 49, 51, 62, 69, 71, 72, 76, 80, 81, 166, 168, 176, 184, 191, 192, 194, 203, 210
farmers, 18, 25, 62, 116, 128, 135, 138, 139, 173, 183
Fass, Paula, 173, 179, 203

gender, xii, xvi, xvii, xviii, 1, 2, 5, 6, 7, 8, 9, 11, 13, 17, 43, 44, 45, 49, 50, 52, 53, 56, 57, 76, 86, 87,

88, 89, 90, 91, 92, 137, 143, 144, 152, 157, 158, 159, 161, 168, 170, 171, 173, 178, 197, 203
generations, xii, xiii, xv, xvii, xviii, 1, 9, 10, 18, 19, 39, 55, 56, 61, 62, 68, 70, 72, 76, 79, 89, 91, 106, 114, 117, 120, 121, 130, 133, 135, 136, 139, 140, 142, 143, 144, 145, 147, 148, 158, 160, 167, 168, 170, 171, 178, 179, 203
Gerald Grant, 2, 4, 149, 152, 178
globalization, xiii, xvi, xvii, 154, 176

hallways, xviii, 31, 51, 74, 92, 105, 106, 114, 141, 202
historical consciousness, 157, 159, 160, 161

identity, 6, 8, 9, 43, 45, 52, 57, 88, 98, 105, 110, 111, 112, 117, 121, 122, 130, 131, 133, 136, 143, 145, 161, 177, 178, 180
immigration, xv, xvii, 4, 175, 176, 180
injustice, 5, 64
institution (institutional, institutionalized), 3, 5, 7, 8, 10, 11, 12, 13, 17, 19, 21, 30, 37, 43, 44, 45, 78, 95, 97, 98, 111, 121, 136, 137, 138, 140, 141, 144, 147, 151, 152, 153, 154, 157, 158, 160, 161, 162, 163, 164, 171
integration, xviii, 1, 2, 3, 4, 5, 6, 8, 10, 11, 12, 13, 36, 37, 38, 52, 61, 145, 147, 152, 177
interracial relationships, 2, 4, 5, 8, 10, 13, 39, 49, 57, 63, 72, 76, 82, 83, 87, 88, 90, 91, 125, 138, 144, 147, 151, 209
intersectionality theory, 6, 157, 159, 160, 161, 163, 173
interviews, 7, 9, 40, 124, 158, 160, 162, 164, 165, 166, 171, 175, 197, 212

Jewish, 4, 105, 107, 108, 109, 112, 113, 122, 126, 134, 136, 203
jocks, 8, 41, 63, 64, 65, 66, 69–74, 76–79, 82, 87, 89, 90, 91, 92, 95, 86, 99, 107, 109, 111, 126, 134, 151, 180, 192, 196

liberal students (liberalization), xv, xvii, 97, 99, 112, 138, 154, 155

Maryland, xii, xvii, 1, 2, 3, 37, 162, 165, 172, 174, 179, 209
memories, xii, xvii, xix, 1, 2, 9, 10, 13, 17, 19, 24, 28, 30, 40, 61, 64, 65, 71, 79, 91, 92, 103, 106, 108, 116, 117, 126, 127, 130, 140, 157, 158, 160, 165, 170, 171, 182
middle-class, xviii, 4, 10, 12, 17, 18, 19–52, 53, 54, 56, 57, 66, 71, 72, 78, 80, 82, 83, 84, 90, 107, 108, 111, 114, 119, 122, 127, 138, 139, 140, 147, 148, 150, 153, 166, 168, 169, 183, 188, 193, 207, 209, 212
multiculturalism, xv, 176
multiracial schools / demographics, 4, 154, 160

nationality, xi, xii, xv, xvi, xvii, 1, 5, 6, 7, 13, 113, 122, 133, 136, 158, 159, 178
No Child Left Behind Act, xvi

oral history, 7, 14, 157, 159, 180, 201, 211, 212, 213

Palumbos, Robert, 3, 179
peer-groups, 87–199
poor students (economic status), xiii, xviii, 1, 5, 8, 10, 11, 12, 13, 19, 21, 23, 24, 25, 26, 28, 29, 31, 33, 38, 43, 44, 46, 47, 48, 49, 50, 53, 54, 55, 56, 62, 64, 67, 69, 70, 74, 78, 79, 81, 83, 84, 85, 96, 97, 117, 122, 148, 150, 151, 166, 168, 186, 207

Portelli, Alessandro, 7, 180, 211
post-school life opportunities, 2, 53
potheads, 63, 74, 76, 77, 78, 79, 82, 84, 86, 89, 92, 96, 97, 105
pregnant girls, 24, 25, 40, 81, 83, 84, 93, 109, 118, 150, 186
preps, 63, 85, 86, 107, 109, 116, 117, 119, 120, 121–123, 126, 127, 133–135, 138, 139, 151, 154, 168, 169, 205, 212
principal, 19, 21, 22, 23, 30, 32, 33, 38, 41, 57, 70, 92, 94, 95, 96, 106, 107, 114, 116, 124, 127, 141, 144, 166, 187, 191, 199, 204
public schools, xv, xvi, xviii, 1, 2, 4, 18, 46, 55, 107, 136, 154, 180, 181, 209, 210
punks, 8, 105, 109, 126, 127, 145, 206

race, xii, xiii, xv, xvi, xviii, 1, 4, 5, 6, 7, 8, 9, 10, 11, 12, 13, 17, 39, 43, 44, 45, 48, 49, 52, 53, 54, 56, 57, 63, 66, 67, 73, 74, 83, 86, 88, 89, 90, 92, 95, 99, 103, 110, 122, 124, 130, 133, 134, 136–140, 144, 148, 150, 151, 152, 154, 158, 159, 161, 168, 170, 173, 174, 177, 178, 179, 180, 181, 182, 193, 194
racism, 3, 27, 57, 87, 115, 125, 130, 143, 144, 151, 152, 153, 207, 210
religion, 5, 6, 7, 13, 103, 112, 113, 122, 133, 136, 158, 159, 178
resegregation, 1, 2, 5, 99, 176, 182, 209, 218
rich / wealthy students, xviii, 1, 8, 13, 18, 29, 43, 44, 45, 46, 48, 50, 55, 64, 66, 68, 69, 83, 119, 122, 123, 133, 139, 168, 187
    see also wealthy students
rural, xviii, 3, 4, 18, 25, 39, 44, 48, 54, 57, 61, 79, 104, 133, 178, 179, 201
rural-suburban, 4, 133, 178

Russian, xiii, xviii, 4, 5, 10, 103, 104, 105, 109, 110–123, 127, 128, 129–134, 135–143, 145, 148, 151, 153, 154, 163, 165, 169, 174, 201, 204, 205, 206, 211

school spirit, 24, 57, 80, 106, 107, 120, 121, 123, 138, 186, 203
segregation, xii, xv, 1, 2, 3, 5, 8, 11, 13, 52, 53, 56, 62, 74, 89, 96, 97, 103, 126, 131, 133, 134, 142, 143, 148, 149, 151, 152, 182, 208, 209, 219
social capital, 83, 85
social categories, 3, 6, 7, 13, 44, 54, 57, 91, 133, 134, 159, 160, 161, 180
social efficiency, 11, 95, 200
social markers of difference, 6, 159
social memory, 158
social reproduction, 54, 210
sports, xviii, 1, 10, 12, 20, 22, 34, 35, 36, 38, 39, 40, 44, 46, 47, 48, 52, 64, 65, 66, 68, 70, 72, 73, 74, 76, 77, 78, 85, 89, 92, 94, 95, 98, 99, 119, 120, 123, 125, 133, 144, 148, 151, 152, 173, 192, 194, 197, 198, 199, 205, 206
student activism, 3, 5, 179, 200
suburban spaces, xvii, xviii, 2, 3, 4, 8, 61, 62, 78, 79, 83, 88, 104, 114, 126, 127, 133, 178, 183, 206, 217

teachers, 5, 6, 8, 12, 13, 17–29, 31, 32, 34, 38–43, 49, 50, 53–57, 61, 62, 67–72, 75–80, 85, 89–97, 99, 106, 108, 111, 113, 127, 129, 135, 138, 139, 140–143, 150, 151, 153, 165, 166, 173, 176, 183, 184, 185, 187, 188, 189, 191, 193, 194, 196, 197, 200, 203, 204, 206, 211, 212
tracks (tracking), 11, 29, 43, 48, 53, 56, 57, 134, 139, 141, 152, 154, 182, 208

academic track, xv, 1, 10, 11, 12, 21, 22, 28, 29, 33–38, 41, 42, 44, 46–48, 53, 54, 56, 87, 95, 97, 114, 122, 133, 140, 141, 148–154, 173, 190, 203, 212

general track, 26, 28, 29, 38, 42, 43, 44, 48, 52, 53, 113, 122, 185, 194, 200

gifted and talented track (GT), 109, 111, 114, 115, 117, 118, 120, 141, 207

honors track, 108, 109, 114, 115, 120, 203, 207

vocational track, 10, 11, 47, 48, 53, 148, 149, 154

transcriptions, 157, 170

urban spaces, xvii, xviii, 3, 4, 8, 9, 97, 104, 129, 133, 177, 178, 210

violence, 10, 35, 38, 79, 92, 114, 115, 122, 139, 142, 144, 148, 150

wealthy students, xviii, 1, 8, 13, 18, 29, 43, 44, 45, 46, 48, 50, 55, 64, 66, 68, 69, 83, 119, 122, 123, 133, 139, 168, 187
see also rich / wealthy students

whites, xii, xiii, xv, xviii, 1, 3–5, 8–13, 17–63, 65, 66, 67, 69, 70, 71, 72–93, 97, 98, 99, 103, 104, 107–116, 121–128, 133–140, 142–153, 159, 162–164

white, lower-class / working class, 10, 24, 27, 28, 43, 56, 84, 85, 96

white, men, 29, 41, 48, 50, 53, 54, 78, 79

white, women, 11, 24, 31, 42, 46, 49, 51, 53, 56, 80, 150, 173, 184, 193

white flight, 3, 62, 179

white poor, 4, 26, 104, 105, 114, 116, 128, 134, 135, 136, 137, 138, 139, 145, 154, 187

working-class students, 10, 21, 24, 27, 28, 38, 43, 46, 56, 84, 85, 96, 97, 133, 139, 147, 166, 167, 168, 169, 207
see also class (economic status), lower-class

yearbooks, 20, 40, 41, 46, 47, 61, 85, 97, 120, 121, 123, 126, 138, 157, 161, 171, 172, 173, 180, 181, 184, 188, 189, 203, 212

zero tolerance policies, 8, 12, 99, 113, 115, 127, 137, 138, 141, 145, 149, 181